"In an era where governments ignore scientific knowledge on the severity of the nature and climate crises in their decisions and politicians outright rejecting such knowledge gain strong popular support, Vetlesen's book *Living with Disappointment in the Face of Environmental Crisis: Resilience, Resistance, and Denial* is a highly timely and sharp analysis of the roots of inaction and denial and possible strategies for resistance."

—**Petter Næss**, *Norwegian University of Life Sciences*

Living with Disappointment in the Face of Environmental Crisis

While the impacts of climate change become increasingly severe, efforts to prevent them suffer one blow after the other, as seen in the rise of far-right populist parties in Western democracies. Why does denialism thrive when blatantly contradicted by the realities before our very eyes, be it wild-fires, floods, drought, and melting glaciers? Should we abandon the assumption that the more solid the knowledge about climate change, the more eager will ordinary people as well as political leaders be to take action?

This book sets out to explain the contradiction witnessed between knowledge and action. Inspired by Clive Hamilton's claim that "denial is due to a surplus of culture rather than a deficit of information", the book critiques the focus on "cognitive disso-nance" in individual agents advocated by climate psychology as well as the individualistic bias in liberal political theory. To get out of the current theoretical as well as political impasse, the author suggests three moves are necessary: from knowledge to first-hand experience, and so to feelings; from the tension within the individual to the social organization of denial; and from the obsession with personal responsibility – nowadays in the guise of building resilience – to exposing the complicity of the culture of neoliberalism in the intimately intertwined crisis of politics and the climate alike.

A highly timely and sharp analysis of the roots of inaction and denial and possible strategies for resistance, the book will appeal to scholars and upper-level students with interests in social, political, and environmental philosophy and psychology; political theory; and environmental studies.

Arne Johan Vetlesen is Professor of Philosophy at the University of Oslo, Norway. He is the author of 30 books, with his most recent book being *Animal Lives and Why They Matter* (Routledge, 2023).

Living with Disappointment in the Face of Environmental Crisis

Resilience, Resistance, and Denial

Arne Johan Vetlesen

Routledge
Taylor & Francis Group
LONDON AND NEW YORK

First published 2026
by Routledge
4 Park Square, Milton Park, Abingdon, Oxon OX14 4RN

and by Routledge
605 Third Avenue, New York, NY 10158

Routledge is an imprint of the Taylor & Francis Group, an informa business

British Library Cataloguing-in-Publication Data
A catalogue record for this book is available from the British Library

ISBN: 978-1-041-11601-1 (hbk)
ISBN: 978-1-041-11602-8 (pbk)
ISBN: 978-1-003-66074-3 (ebk)

DOI: 10.4324/9781003660743

Typeset in Times New Roman
by SPi Technologies India Pvt Ltd (Straive)

For Thomas Hylland Eriksen
1962–2024

Content

Preface

This book follows the Routledge trilogy – comprising *The Denial of Nature, Cosmologies of the Anthropocene* and *Animal Lives and Why They Matter* – where I address the natural world and our relationships with it from various angles, ranging from philosophy over anthropology to ethnology. Having strayed from my original training and interests in philosophy, I now return to it, engaging with sociology and psychology in exploring how contemporary society responds to environmental crisis.

My present aim is to employ this multidisciplinary approach to an issue that traditionally has been a hallmark of philosophical inquiry, namely the relationship between knowledge and action, between what we know and how we act. This issue can be traced back to the birth of philosophy Western style, articulated in the question Socrates asked, considering it to be fundamental to ethics and politics alike: Can I know what is the right thing to do, and yet not do it? If I fail in the sphere of action, does that mean that I have gotten it wrong in terms of knowledge and understanding?

Over the centuries, various brands have been used to characterize Socrates's way of framing the issue, most notably "intellectualism" and "idealism". For my purposes in what follows, another term strikes me as more appropriate: epistemic optimism. Such optimism typically plays out as the assumption that the more knowledgeable we are about a pressing problem, helping us recognize it as such, the more readily will we act to solve it. By the same token, the more urgent a problem appears to us by way of the relevant facts, the more motivated we will be to change our practices if that is called for.

Contemporary environmental crisis being this book's case in point, it is hardly controversial to say that here we face a problem that proves the said assumption wrong: the science is crystal clear, the knowledge has been widely available for decades in so-called "liberal democracies", and yet the action called for to prevent a catastrophic outcome is not being taken. Instead of the positive relationship Socrates articulated, plausibly so to generations of philosophers and essential to the idea and ambition of

"Enlightenment" on which Western culture prides itself, recent years have witnessed a reversal: the more serious the problem gets, the more potently defiance and denial are being propelled, preventing even long overdue and modest measures to address it from being implemented.

Why is that? How to explain it? These are the key questions that this book seeks to explore, not by looking for a definitive answer in one particular scholarly discipline, but by engaging with different approaches in the social sciences as well as in philosophy. To anticipate my argument, we shall see that we, both as individuals and as a society, are far less rational than we like to think; less rational, also, than a quick fix, a rapid solution to a crisis that is cumulative, growing as we speak, would call for. After decades of barking up the wrong tree, of looking in the wrong direction, we need to shift the emphasis from knowledge to emotions, from dispositions in individuals to patterns of culture, from the intrapsychic operation of denial to its social organization, making the quest for truth ethically speaking, and for speaking truth to power politically speaking, so much more difficult than the optimistic view about knowledge translating into action suggests. Difficult, yes, but not impossible, not to be given up.

I dedicate this book to Thomas Hylland Eriksen. Thomas, a renowned professor of social anthropology at the University of Oslo, was a cherished conversation partner of mine ever since we studied together in the early 1980s, a public intellectual indispensable as well as irreplaceable in troubled times, a kindred spirit committed to the interplay between, and common threats against, the twin diversities, the cultural and the biological.

My gratitude also goes to the persons who took the time to comment on my manuscript, either in whole or in part: Jan-Olav Henriksen, Per Bjørn Foros, Zemir Popovac, Espen Hammer, Jack Palmer, and Jonas Kittelsen. I have also benefited from conversations about this book's topic with Joachim Renn, Odin Lysaker, Sigurd Hverven, Sindre Brennhagen, Rasmus Willig, Per Nortvedt, Linn Stalsberg, Erik Stänicke, Erling E. Guldbrandsen, Knut Ivar Bjørlykhaug, Solrun Samnøy, Ole Jacob Madsen, and Andreas Malm.

Horten, Norway

Introduction

"Drill, baby, drill"

Recent years have seen the emergence of a correlation that has taken many by surprise: the more visible the consequences of climate change, the greater the popular support for political parties that promise to "drill, baby, drill". Donald Trump's successful bid for a second term as US President may well be the most prominent case in point, but it is part of a broader trend that includes my own country, Norway. The message seems to be that the more indisputable the warnings of climate science turn out to be before our eyes, meaning here and now, not elsewhere and in the future, the more stubbornly is the present trajectory, that of fossil fuels-based economic growth, held on to.

To point out the correlation between two phenomena is not to explain anything; it is to invite questions about why they occur, either simultaneously or in a sequence, questions about cause and effect. What is the correlation I mentioned essentially about?

It is about danger, and responses to that danger. Climate change poses a danger to human society and to non-human species and lifeforms, and as such it requires an adequate response, one that is suited to minimize or eliminate the danger by targeting its root causes. Given this science-based way of putting it, it is noteworthy that such a response is not forthcoming but conspicuously lacking. Shouldn't we expect that the greater and more tangible the danger we are facing, the greater will be our readiness to do what it takes to avert it? And shouldn't this expectation apply *especially* in a situation where what causes the danger are practices that we humans are involved in, indeed help sustain? If the problem is caused by us rather than by some alien force outside our power and control, doesn't this human-caused dimension strengthen the prospect that we will indeed do what is necessary, namely by changing our practices?

Sensible as this way of asking the questions may seem, it is not true to how we humans tick, at least not as members of contemporary society. That said, who exactly are "we" in the above statements? Do all humans respond in like manner? Are all equally "involved" in the practices driving climate change,

DOI: 10.4324/9781003660743-1

destroying habitats, polluting the air and the seas? Who, exactly, need to change their practices, and how? Indeed, it attests to historical ignorance to speak of a unified agent – or victim – here. To be precise, we must speak about the North and the South, the rich and the poor; about geography, colonial history and the class-based society that is inseparable from capitalism as we know it.

This is true: there is no validity, no innocence, to the simple "we" used above; it conceals the very differences – inequalities – between, and within, actors and victims that desperately need to be explicitly stated and sorted out for the required changes in human practices to happen, and for the changes to be in keeping with climate justice. Unfortunately, the factuality of these inequalities in terms of power and historical responsibility is also a major obstacle to reaching agreement about who is required to do what at one COP-meeting after the other. The pattern is well-known since these summits started in 1990: the rich countries are the most responsible for causing climate change, in the past as well as in the present, are not willing to accept having to lead the way in terms of reducing emissions, let alone question the ambition to continue growing their economies – the very ambition that countries like China and India (making up one-third of the current world population) regard themselves as entitled to realize in the coming decades. Add to this the unholy alliance between fossil fuel companies hell-bent on increasing their profits and the billions of people aspiring to adopt the "American way of life" as facilitated by said fossil companies.

Given this gloomy state of affairs, what to do? Where to turn to get a better grasp of why the danger so clearly posed by environmental crisis is growing with every new report and update issued, instead of being reduced? How to explain the fact that the popular support for ambitious environmental policy goals was stronger in the past – say, five years ago – than it is today, when the problem objectively speaking is even more serious and alarming? Again, why is the correlation between danger and willingness to act to avert that danger of the negative kind we are witnessing?

I readily admit that in stating the issues like this, I employ the oversimplified "we" that I criticized above. Perhaps stylistically there is no way to avoid this, or to do so whenever necessary for the sake of nuance and precision. In the chapters that follow I do draw on some studies that subscribe to the (largely tacit) premise that human psychology can be described in a generalized, once-and-for all manner, applicable irrespective of the particular region, history, and culture of the individuals in question – Freud's analysis of the mechanisms of repression is a case in point. Sometimes such a general approach, with its sweeping claims, is both warranted and unavoidable. Even so, when climate change and responses to it are the topic, I consider it vital to state clearly who we are talking about: what sort of individuals, with what sort of history, and impacted by what kind of cultural values, norms, and expectations? Climate change, or environmental crisis more broadly, is precisely *not* an issue, *not* an occurrence and process that takes place *sub specie aeternitatis*, but is instead eminently situated in time and place, culturally and socially as well as in the natural world. That some of the most influential scholarship devoted to this topic fails to heed this, is a critical point I shall be making time and time again.

On17 May 2024, *The Guardian Weekly* ran a cover story titled "We live in an age of fools", based on interviews with 380 climate scientists from all over the world. Typical quotes are "I am scared, I don't see how we can get out of this mess", "I am hopeless and broken", and "I worry about the future my children are inheriting" (Carrington 2024: 34). Key findings are that 77% of respondents believe global temperatures will reach at least 2, 5 C above pre-industrial levels, "a devastating degree of heating"; almost half – 42% – think it will be more than 3 C; and only 6% think that the 1, 5 C limit (the ambition agreed as part of the 2015 Paris Accord) will be achieved. Maisa Rojas, an IPCC scientist and Chile's environmental minister, says: "We need to communicate that acting on climate change can be a benefit, with proper support from the state, instead of a personal burden". Another scientist, Elena Lopez-Gunn, at the research company Icatalist in Spain, states that "I believe in social tipping points, where small changes in society trigger large-scale climate action", adding "unfortunately, I also believe in physical climate tipping points".

A couple of months later, on 2 August 2024, *The Weekly* ran another cover story on climate change, this time focusing on how the rich countries most responsible for creating, and exacerbating, the problem are stepping up – not to combat it, by reducing emissions, but by *increasing* the volume of emissions: 825 oil and gas licenses have recently been handed out by countries like the United States, China, Canada, Russia, and Saudi Arabia. Norway is part of this club, projected to hand out 80 oil and gas licenses during 2024, resulting in 771 million tonnes of greenhouse gases pollution – the equivalent of putting 183 million new gasoline-powered cars on the road. These facts make for an outrages contrast with the message issued by the International Energy Agency (IEA), namely that no new oil and gas project can proceed if the Paris agreement's goals are to be met (see Milman 2024: 12; Beckert 2024).

Consider these statements and facts in light of what the Australian philosopher Clive Hamilton writes in his seminal article "What history can teach us about climate change denial", published in 2013:

> In these circumstances, facts quail before beliefs, and there is something poignant about scientists who continue to adhere to the idea that people repudiate climate science because they suffer from inadequacy of information. In fact, *denial is due to a surplus of culture rather than a deficit of information.*
>
> (Hamilton 2013: 17; my italics)

What does Hamilton mean when he says that "denial is due to a surplus of culture"? I think that this claim makes more sense today than it did when Hamilton wrote his article more than ten years ago. At that time, the empirical evidence he had in mind, most prominently in the United States but also in Western Europe and in his native Australia, was to do with "the sense of a nation in decline, the fragility of liberal forces and the rise of an angry populist right" (2013: 21). This sounds all too familiar today, but note that Hamilton addressed a world that had not yet seen the potency of the Tea Party movement, let alone the emergence and eventual triumph of Trumpian MAGA-policies, inspiring in turn the electoral breakthrough of kindred spirits and parties on the far right throughout Europe, to which must be added the likes of Bolsonaro in Brazil and Milei in Argentina. "Environmental policy and science", Hamilton points out, "have become battlegrounds in a deep ideological divide that emerged as a backlash against the gains of the social movements of the 1960s and 1970s" (2013).

This also sounds right today, albeit with the corrective that a couple of years after Hamilton's analysis many of the countries mentioned saw the unexpected rise of an environmental movement, Fridays for Future, led by a teenager from Sweden, Greta Thunberg, said to be the largest political movement made up of young people ever seen, both in numbers and in global reach. Indeed, in 2019 and 2020, when thousands of people took part in protests organized by Extinction Rebellion in countries like United Kingdom and Germany, it looked as if the backlash Hamilton so succinctly associated with the rise of a populist right promising to put an end to policies addressing climate change was itself in for a serious setback: environmentalism having suffered a blow, it seemed that the time had come to bounce back and take a big step forward.

The reason why Hamilton's diagnosis still holds, notwithstanding the mobilization just mentioned, is that it suffered what can only be called a sudden death: the COVID-19 pandemic and the lockdown it was met with meant that those who had taken to the street to call for climate action could do so no longer. Two years later, when the lockdown was lifted, the movement proved to have lost its momentum, its historical moment, all but completely. And as we will see in Chapter 4, on resistance, the fewer the protestors, the more brutal the measures deployed to intimidate and punish them carried out by the police, the courts, and the governments in the "liberal democracies" in question. As for protests in countries like China, India, and Russia, so much a part of the problem of climate change, they are strictly forbidden, whereas Brazil holds on to its long-time record as the country where the largest number of activists are killed, with impunity at that.

Despite this back and forth among the political winds blowing during the last couple of decades, one factor is constant, and it is the one I take Hamilton to refer to in his first statement, even though he doesn't name it:

the culture of neoliberalism. This is a culture, a mindset, a way of looking upon and organizing society and the relationship between individuals and institutions that has been dominant in Western societies since the electoral triumphs of Reagan and Thatcher, and its history has been told many times (see Harvey 2005; Monbiot & Hutchison 2024).

Since for most people living today neoliberalism is the only culture they have experienced, it should not be necessary to set out its characteristics here. Plausible as this claim may seem, there are good reasons to turn it around: *precisely* because that kind of cultural environment is the only one we have experienced, we tend to naturalize it, that is, to take it as the "natural" way for a society to be understood and organized, implying such diverse meanings as "rational", "necessary", "inevitable", and "desirable". Specifically, the content of what is thus internalized by each of us as "the way the world works" gets defined *for* us – not by us qua individuals – as what is meant by such values as freedom, responsibility, entitlement, and justice; what I and others deserve, particularly in terms of consumption and possession of material goods; what makes some individuals and groups more deserving than others; what constitutes virtues to be promoted and vices to be avoided; what part the human condition of vulnerability and dependence shall play; and last but not least, how to make the distinction between experiences and feelings suited to be shared with others, and those not to be expressed, or not felt at all.

More specifically, we shall see that the cult of personal responsibility, with its emphasis on self-optimalization and self-sufficiency, is at the heart of the neoliberal culture's demands on individual performance. This overall *shift from collective to individual agency* alters the constellation between institutions in general, the (federal) government and the state in particular, on the one hand, and the individual on the other. The decades since the late 1970s in Western liberal democracies such as the United States, the United Kingdom, and Norway have witnessed a shift from a situation where institutions were set up to compensate for the dependencies of needy individuals by securing for them what they cannot muster themselves (the welfare state as we used to know it), to one where the relationship is *turned around*, so that now individuals are expected to compensate for the ever-diminishing support and security meted out by institutions which have become marketized, fixated on "efficiency" and budget discipline, hence less and less distinguishable from corporations run by profit, the private sector and its logic having become the model to be adopted by the public sector as well. Brought to bear on present day "crisis management", exacerbated by climate change, and by the recent global COVID-19 pandemic, the macrolevel shift just described plays out at the microlevel in the form of a rising demand for building "resilience", not in a communal sense, but strictly individualized.

To be sure, the above may seem a theoretical list of changes. But in fact it is utterly concrete, because it covers the way we actually live under the cultural hegemony of neoliberalism. And as we shall see in greater detail below: if we venture to *not* live like that, to communicate concerns and express feelings we are not – culturally – supposed to do, there will be a backlash; such behaviour will not go unnoticed and without feedback, ranging from outright hatred and aggression to studied indifference and stubborn silencing. Granted, this is not the type of backlash Hamilton talks about, articulated on a pronounced political level and pinpointed by what kind of parties win or lose elections, what social movements are currently on the rise or losing momentum. The dynamic I have in mind is more subtle than that, not forthcoming in the form of surveys or results on election day. Its arena is the interaction between individuals, typically – for my topic – that involving a person who is worried about environmental crisis and what it means for the future, and another who acknowledges no such worries at all, considering them groundless – or perhaps fearing that they are indeed, to some extent, valid, yet being unwilling to entertain that possibility and the consequences it would carry. The model for the approach I am referring to, and will develop as the book progresses, is inspired by the sociologist Erving Goffman, the author of the modern classics *Asylums* and *The Presentation of Self in Everyday Life*.

Make no mistake: I am not suggesting that a micro-perspective take precedence over a macro-perspective. Indeed, to enact such a primacy that of individual agent over structure and institutions, is at the heart of what a neoliberal culture is all about, what it induces us to do, as scholars and laymen alike. To avoid complying with this expectation, I shall endeavour to move between individuals interacting with each other one-to-one, and individual agents as they are part of groups and relate to institutions of various kinds, including the government and the state. A crucial feature of neoliberal culture is to do with the latter relationship, instructing individuals to see themselves first and foremost as consumers, competing with others over scarce goods, and being at fault – qua individual – whenever one turns out a loser, and deserving the success whenever a winner. Here the role of institutions is largely bracketed, and the state is viewed not as an ally to count on when needy – to allude to what sociologist Zygmunt Bauman calls "the social state" – but with distrust and suspicion; a negative view that we will see is not only a key feature of neoliberalism itself, but that – surprisingly – has been adopted by many of its prominent critics on the left as well, as I shall argue when discussing Michel Foucault, Judith Butler, and (in the context of civil disobedience and activism) Andreas Malm below.

Having made the observation that climate change denial "is due to a surplus of culture rather than a deficit of information", how does Hamilton go on to describe that culture? The short answer is that he doesn't. Or to be more precise, he says very little about it, restricting himself to statements like this: "In the most vital test of our capacity to protect the future through the deployment of rationality and well-informed foresight, the 'rational animal' is manifestly failing" (2013: 28). He adds:

> The 'autocratic subject' could extract knowledge from nature but could also choose to ignore that knowledge if it unsettled the mind. It was a double-edged subjectivity that had the self-certainty both to liberate objective science and reject the facts it uncovered when convenient. The climate crisis is upon us because we are intoxicated by our subjectivity
> (2013)

What does it mean to be "intoxicated by our own subjectivity"? Again, Hamilton stops short of elaboration. But I think he is onto something very important about what sets a neoliberal culture apart from its predecessors in Western history. For a start, the prestige accorded to knowledge as attained and – in principle – made public thanks to science, helping to "enlighten" society at large about the consequences of advances in knowledge, may always have been in large part a matter of "ideology" or even "myth" (to allude to Horkheimer and Adorno, arguing to this effect in their book *Dialectic of Enlightenment*). Yet something deeper and more unsettling has occurred with the triumph of neoliberalism: it fosters a mindset that abhors the science-based knowledge that Hamilton refers to when recalling the legacy of the Enlightenment; ours is increasingly a society that puts those "in the have" (typically measured in money and power, perceived as inseparable) way above those "in the know", associating the professions dedicated to science with an elite out of touch with ordinary people, arrogant, and aloof. There is clearly a class division to this, as we shall see when turning to the work of Arlie Russell Hochschild below. Regrettably, I cannot do justice to the significance of class in this book, except for commenting on it when discussing the Marxist position of Andreas Malm in Chapter 4.

However, to speak about the knowledge generated by science as being simply rejected by a majority of people is both too strong and too crude. We are encouraged to admire the science behind Musk's Tesla and his spaceships, and to be optimistic, albeit it with some trepidation and ambivalence, about the prospects of full-scale implementation of AI in various professions and workplaces. Indeed, on the whole, technology is a secular god in itself, a favourite fetish, especially among men, and belief in its potential – be it to "solve" the climate crisis, say, by way of geoengineering – is widely shared (see Hamilton 2014). Yet the widespread admiration for the wet dream of going to Mars championed by billionaires like Musk and Bezos cannot hide the fact that while this dream is one born of a wish to *escape* the perils of climate crisis, such escape is beyond the reach of the 99% doomed to suffer that crisis on this one planet that we are stuck with. On the other hand, and this is my claim, we are discouraged from taking the science-based knowledge about the causes and consequences of climate change too seriously, or only up to a point, by "taking note" cognitively speaking, but not seriously in terms of our experiences and feelings. The latter dimension, the lived as distinct from the cerebral one, the existential as opposed to academic one, is a no-go area, a non-starter socially speaking. And to the extent that you, nonetheless, break the cultural taboo of venturing there, trying to share doing so with others, chances are you may precisely *not* bond with them over the issue – climate change as real, as here and now, corroborated by a science whose findings are truly alarming – and what it does to you as a person, but also as a citizen, a political being.

Again, if this sounds overly theoretical, much as I make a plea to access the dimension of lived, first-hand experience, it will get less so once we turn to the person-based case studies in Chapter 2, where the stories of Sally, an American woman, as told by psychoanalyst Renee Lertzman, and Eirik, a Norwegian man, as told by sociologist Kari Marie Norgaard, will allow us to delve deeply into how difficult it is to "live with" the knowledge of ongoing climate change, of worsening environmental crisis and its impacts on life in your home place.

But again, we need to be more exact about what is at stake. The difficulty, I submit, is not that of living with the knowledge; it is that of coping with the *experience* of climate change, of a natural world that is changing, locally as well as globally, meaning *ever closer to home*, visibly and tangibly so. If anything is different today from the situation Hamilton was describing in 2013, it is that things – the realities on the ground, in the air and in the seas – have moved from knowledge in the academic sense, often based on scientists' estimations of future scenarios (typically, how much will temperatures rise by 2030, 2050, and 2100?), to first-hand experience in the present tense, that of here, now, where you and I live.

This is important for the phenomenon so crucial to Hamilton and to me alike, that of denial. What is being denied when we talk about denial in the midst of environmental crisis? What goes to make up its sources, its objects, its targets; and what are its root causes? If we come up with different answers for the different factors listed; and if there is a *mismatch* rather than match between root causes and targets, say, between a growth-obsessed economy on the one hand, and the aggressive attention directed to scapegoats on the other, then denial starts to become a truly complex phenomenon, not only academically but with regard to how such denial is lived, acted upon, and shared with others in the community, others sharing – and that will remain absolutely crucial throughout – the predicament of living in the heyday of a neoliberal culture, as this holds for the material I will discuss in the book, taken, by and large, from contemporary society in the United States and in Norway.

To engage in denial about what I first-hand *experience* is different from denying a piece of *knowledge*, even though in both cases the endeavour to engage in denial is made easier by others' complying, that is, by engaging in the same act(s) of denial, rendering it what *we* do, how *we* respond, thereby taking part in normalizing it, reproducing it as a shared social practice. Indeed, it will be a major task in this book to show precisely how denial is socially organized (which is deeper than simply shared, because structural), whereby the pervasiveness of organizing it thus is corroborated by a culture that favours, or even commands, such denial to take place, to be sustained and even amplified over time – demanding, it turn out, an ever-greater effort the more undeniable the experience of climate change and a changing natural world qua real becomes.

Let us never forget that *denial is hard work*, even when shared with so many others, be it the majority of people we interact with (die-hard activists exempted, forming a subculture of their own); just how hard it is will become clear when we turn to the case stories below. In his classic article "Repression", Freud writes that "the essence of repression lies simply in turning something away, and keeping it at a distance, from the conscious" (1991 [1915]: 147). While "*individual* in its operation", repression is "also exceedingly *mobile*". Freud continues: "The process of repression is not to be regarded as an event which takes place *once*, the results of which are permanent as when some living thing has been killed and from that time onward is dead." Rather, he explains, "repression demands a persistent expenditure of force, and if this were to cease the success of the repression would be jeopardized, so that a fresh act of repression would be necessary" (150f.; italics in original). Elsewhere Freud talks about "replacement of external by psychical reality" (191).

Note that repression and denial are not the same; they are not identical phenomena. Repression is one of several tools employed for the mechanism of denial ("Verleugnung" in German) to be effective, and Freud's overall understanding of their interplay is shaped by the role he attributes to them in an individual's ongoing "pleasure–unpleasure regulation", put starkly in the claim that "the motive and purpose of repression is nothing else than the avoidance of unpleasure" (153). The crude reductionism Freud subscribes to here has been pointed out by many commentators, and often duly criticized.

I mention Freud's original statement about repression because it sheds light on its *modus operandi*, its having to be repeated, again and again, demanding "a persistent expenditure of force". Even though, as we saw, Freud showcases repression – or denial in my case – as something that is "individual" in its operation, it is nevertheless, as I noted above, an endeavour that may well be sustained – aided, affirmed, cemented – by that individual's significant others, or by society at large, rendering the expenditure of force a communal undertaking and it that sense less exhausting for the individual who is "in the midst of it", its subject and object alike. Denial is deeply and lastingly *socially contagious*, especially when engaging in it becomes part and parcel of socialization and identity formation, of proving oneself as "one of us", as is indeed the case – so my claim – in the current neoliberal culture. The costs of breaking out, of refusing to be part of denial and its constant reproduction, are considerable, too costly for many of us. This will be a major topic in Chapters 3 and 4.

There is another aspect to Hamilton's somewhat dated emphasis on *knowledge about* climate change, brought out when I pointed out the shift towards actually *experiencing its reality* here and now, say, in the form of floods, to pick an example that applies to countries as otherwise different and far apart as Pakistan, China, Spain, and Norway, to say nothing about regions affected, often meaning devastated, by typhoons, by hurricanes, by wildfires, by drought, and by melting glaciers. If anything, scientists have *underestimated*, not alarmism-like exaggerated, the accelerating pace and scope of the ongoing changes and how they impact societies as well as the natural world they are part of and remain dependent upon.

What does it mean, this shift occurring during the last decade or so from climate change being an issue of knowledge to it becoming increasingly one of experience? Is the one more, or less, contestable, disputable, deniable than the other? Perhaps there is an academic answer to this. But perhaps, more importantly, the question is the wrong one, mistakenly posed. Why? Because it – still – gives privilege to knowledge, to a certain, verifiable kind of knowledge, whereas the significance of the shift, if we take it seriously as in fact occurring, demands that we shift focus, too, in our attention. What do we need to pay closer attention to?

In her book *Stolen Pride: Loss, Shame, and the Rise of the Right* (2024), the American veteran sociologist Arlie Russell Hochschild, born in 1940, continues where she left off in her acclaimed *Strangers in Their Own Land* (2016), by exploring the cultural narrative that is crucial to the popular support for Trump's brand of far right policies. Shifting her field work from the rural population of Louisiana to people living in Kentucky, Hochschild's objective remains the same: to understand the relationship between the three key emotions she considers instrumental in the rise of the right, namely pride, shame, and blame. Her finding is that changes in how these emotions relate to each other, in who owns them, whom to trace them to and whom to direct them at, is vital to understanding the radical changes in the political landscape during the last decades.

Hochschild's premise is that we typically feel eager for pride, and angry for being shamed. The kind of pride she encounters in her subjects is not to do with arrogance or conceit, but of "being of use", and thereby gaining honour, respect, and status from others, making for solid self-esteem. Shame is the mirror opposite of pride thus experienced, referring to feelings such as humiliation, mortification, or embarrassment; shame is "felt as an unpleasant sense of self-deflation", coming with a sense that I have done wrong in the eyes of others (2024: 26). We see that the distinction between pride and shame is evaluative in that the former is sought for and the latter avoided. Shame threatens to spoil and undercut pride.

What gives rise to these key emotions – when to feel them, to show them or to hide them – is a matter of the culture one belongs to, and not for the individual to decide. But Hochschild's interest is a more specific one: namely individuals' placement in "the pride economy": "We are born into a region, a social class, a race, a gender, and these increase or decrease our value in the wider pride economy" (26). She gives an example well-chosen for her topic: "Pride in the job of a coal miner and in the moral fortitude and know-how the job requires, pride in being in a region so central to the nation – all these have shifted" (26).

The nature and consequences of the multidimensional shift in question take up the rest of Hochschild's book. For my purposes, the shifts to be highlighted are the following: pride is felt to be "stolen", its sources devalued; lost pride gives rise to shame, undermining identity and self-esteem; such shame needs to be gotten rid of, and an effective way to do so is by diverting it to others, by way of blame; blaming others, viewing them as responsible for lost pride, or as an obstacle to regain it, allows for a legitimate venting of anger: they get what they deserve.

This sequence – from pride over shame to blame and anger – is recurrently alluded to, implicitly or explicitly, in the conversations Hochschild has with various sorts of people, across group-based divides. A section where she paraphrases "the message" of Matthew Heimbach, a white nationalist planning to lead a march in Pikeville, Kentucky, goes like this:

> We hear your distress. There's much you do not have. We're here to raise the value of what you do have – your whiteness. [...] You've lost your regional pride, your well-paid jobs, suffered devaluation of what you do have, and you've had enough. We on the violent right will erase shame from you and seamlessly divert your shame to blame – blame of Jews, Muslims, Blacks, immigrants, liberals, and Democrats. Your access to the American Dream? They took it!
>
> (2024: 45)

In this rhetoric and the cultural narrative it helps sustain, taking it in an increasingly radical far right direction, "stolen" becomes the key notion, setting the above-mentioned sequence of emotions into motion. Common to all cases invoked is that "stolen" refers to *the right to something gone*, something good that has been spoiled, destroyed, or taken away. To say that something is gone, or about to be gone, is to imply that it was or is presently being stolen, raising the question of who did it. Indeed, to translate and transform the experience of loss into an accusation of stealing is to politically weaponize it. Applying her analysis to Trump, Hochschild observes that "through his daily rhetoric of victimhood, Trump guided the emotional needle from 'loss' to 'stolen'. Over time, it grew into a master narrative, and like a magnet, it gathered further losses to the idea of 'stolen'" (2024: 223). The ever-expanding list of things "stolen" reads as follows: "Election; Appalachian land; proud region; good jobs; community; story of heroic America; an undisputed concept of human sexuality; honourable country culture; white power; recognition of one's struggles and accomplishments; visibility itself; right to guns; pride" (223f.).

Given the topics that I shall examine in the chapters to follow, several things are worth pointing out in the light of Hochshild's analysis.

First, despite the sheer number of things – or goods – "stolen" just listed, aren't some crucial items lacking? What about *a flourishing natural world*, including the very basics, clean air, clean water, and a fertile soil? And what about *a viable future*, faith in the notion that things will get better? As for things gone, or disappearing, or becoming few and far between, we may mention wildlife and animals that until recently used to be part of the environment, in Kentucky no less than in Louisiana, including a large number of fish and birds. What is gone, and in that sense qualifies for being "stolen", or taken away, making for a substantial addition to the many things listed in the quote from Hochschild, is the good of – and should we say right to? – being able to see and listen to a rich variety of animals and birds; to go swimming in the sea or the river or the pond; to go fishing and hunting in one's home area, being part of rural America, that is to say, a country and a region where all of these things were available, common and within reach, part of one's way of life, and in that sense, of one's *pride* – there it comes – in the home place and its qualities, thus also indispensable to one's sense of identity as living in that place, enjoying a sense of belonging to it, experiencing and doing all these things, all of which – in my additional list – are part of the natural world as it plays out in the place in question. These are not minor or peripheral changes, or marginal sources of loss; they are eminently concrete, first-hand experienced, rich in social and cultural significance, and for the quality of life of the community as such. The changes-cum-losses play out utterly concretely as places you can no longer go to, activities you can no longer indulge in, practices and skills you can no longer uphold, let alone pass on to children or grandchildren. An enormous impoverishment of human, social, and cultural life is coming to pass as part of this impoverishment of the natural world that provides its framework. Inextricably intertwined, cultural diversity and biological diversity are endangered by the same root causes, as Thomas Hylland Eriksen shows (2024).

Yet this entire dimension of loss that affects the people Hochschild interviews and the places they call home makes for a conspicuous unsaid in her books. True, this is somewhat less the case with *Strangers in Their Own Land* than with *Stolen Pride*, and I shall say more about the former at the end of Chapter 3. And to be fair, Hochschild at one point does mention the environmental impact of so-called "mountaintop removal", where "tons of soil and rock are bulldozed down the sides"; "it contaminates water, disrupts plant and animal life"; "it makes valleys vulnerable to flooding, and creates acres of bare rubble which becomes graveyards of their own" (2024: 260).

But why, one may ask, should Hochschild lead her conversations in this direction if the people whose stories about lost pride and about shame and anger over things "stolen" do not themselves bring it up? Isn't Hochschild's selectivity – privileging the cultural and man-made over the natural and organic, economy over ecology, as it were – merely a correct, if regrettable, echo of a factual *disconnect?* Namely the disconnect between the present-day human lifeform and the natural world within which it is situated, whereby urbanisation, new technologies and the like render nature qua first-hand experienced (seen, heard, smelled, touched, and lived-in), qua meaningful and a source of identity and belonging, more abstract and experientially remote for every decade, every year passing, with so many activities having shifted in two generations' time from the outdoors to the indoors – as a result, intended or not, of "progress" and "development"? In short, isn't the bracketing and the silencing of nature in this sense, this cultural and mental as well as physical "turn inward", away from the outside world and its countless multispecies encounters, and amplified by technological innovations like smartphones, a simple fact, like it or not?

This is a good question. But perhaps not, or so I shall argue, for the reasons assumed in finding it so. By not paying attention to the "nature" dimension of the losses impacting the life of those affected, in effect one-sidedly attending to, tapping into and lending voice to the "culture" dimension, Hochschild not only misses out on the former in its own right – she also misses out on the inter-connection between the two. Such analytic impoverishment conceals the factual ongoing one that needs to be illuminated and understood.

The costs of this double missing out are considerable. For a start, to employ the distinction pioneered by Erving Goffman (1959), it means that Hochschild tends to throw all the light at what is being said "frontstage", taken to be proper for such exposure and sharing, thereby allowing the "backstage" to remain in the dark, meaning the thoughts and feelings, concerns and anxieties that the individuals involved regard as "unsuited" to being displayed and expressed to others. Indeed, as Hochschild herself shows so clearly, *what* persons are ready to perceive and name as the causes as well as targets of their key emotions – pride, shame, anger – is eminently culturally shaped, giving the green light as it were to one category of causes and targets, the red light to others, keeping them in the social and psychical dark. To pick a case in point, even the obvious ambivalence of coal, namely as on the one hand a source of people's livelihood and thus of pride, and on the other a cause of environmental degradation, ill-health and loss of quality of life for humans and nature alike, fails to be explicitly addressed. Hence Hochschild risks reinforcing, rather than critically questioning, the said culturally induced and socially upheld suppression of the entire dimension at issue here. Again, this does not only make for an analytic lacuna, though that may be important enough. Existentially, the suppression – qua internalized social expectation and cultural norm – renders people speechless, both metaphorically, symbolically, and literally so, depression frequently being a psychological consequence, isolation and loneliness a social one. Little or nothing of this comes to the fore in Hochschild's conversations; plenty of it does in the studies by Lertzman and Norgaard we shall examine in the chapters below.

If this seems an unfair criticism of Hochschild's book, let me bring in a point closer to her stated objectives – namely, that there exists a connection between *what people feel shame about and what they are in denial about*. To the very extent that we do attend to what is being denied, it goes without saying that it does not appear "frontstage" but is hidden "backstage"; Freud would say underneath, down under, alluding to the unconscious, his metaphor being vertical as opposed to horizontal in Goffman's distinction. By omitting any systematic interest in denial from her approach, privileging the said over the unsaid almost completely, *unsaid because culturally unsayable*, Hochschild not only ends up with a less complex understanding of what the changes her subjects are affected by mean to them. She also, to make the shift from micro to macro, individuals to structure, leaves unaddressed a fundamental fact, making it the elephant in the room: that the *economy* that these people in contemporary American society depend on – for their living as well as their pride – in its turn depends on the ongoing degradation and exhaustion of the natural world which that economy is part of and remains dependent upon, like it or not; and that the *culture* they belong to does its utmost to make sure that this very connection, the exploitation of human capital and that of nature as two aspects of the same dynamics, is unsaid and unacknowledged.

It may well be that if Hochschild had tried to probe deeper, to encourage her subjects to speak not only about what it turns out they want to speak about, she would have found that it would "kill the conversation" – this being what both Lertzman and Norgaard experience when raising questions about living with real time climate change in a rapidly changing natural environment. However, that some topics turn out to kill the conversation makes them all the more intriguing, inviting questions as to why and with what consequences for people who would much rather talk about something else.

As signalled by its title, there is an affinity between this book and Christopher Lasch's *The Culture of Narcissism: American Life in an Age of Diminishing Expectations*, published in 1979. One of the virtues of Lasch's book that makes it a rewarding read even today is his insistence that psychoanalytic theory helps enrich sociological studies and the critique of culture; so crucially so that the latter are all the poorer without it. Specifically, Lasch invokes the notion of "maturity" as indispensable for the project of critically exploring contemporary society and its ills, its lack of sanity, its potential for irrationality and self-destruction. These are big labels, or slogans if you like, that virtually nobody will fault a present-day sociologist – say, Hochschild – for doing without, deeming it a good thing that the "moralizing" and "patronizing" sweeping claims of *Zivilisationskritik* á la the early Frankfurt School (Horkheimer, Adorno, Marcuse, and Fromm) are long gone and today totally out of academic fashion.

Yet something is lost along the way. A couple of quotes from Lasch makes it clear what I am getting it. In his Afterword, he writes:

> The best hope of emotional maturity…lies in acceptance of our limits. The world does not exist merely to satisfy our own desires; it is a world in which we can find pleasure and meaning, once we understand that others too have a right to these goods. Psychoanalysis confirms the ancient religious insight that the only way to achieve happiness is to accept limitations in a spirit of gratitude and contrition instead of attempting to annul those limitations or bitterly resenting them. […] Technology expresses a collective revolt against the limitations of the human condition. It appeals to the residual belief that we can bend the world to our desires, harness nature to our own purposes, and achieve a state of complete self-sufficiency. […] Nature retains the upper hand: The very technologies designed to overcome natural limitations on human comfort and freedom may destroy the ozone layer, create a greenhouse effect, and make the Earth unfit for human habitation.
>
> (Lasch 1990 [1979]: 242, 243f)

The British psychoanalyst Sally Weintrobe, a pioneer in making that tradition fruitful for understanding the *Psychological Roots of the Climate Crisis* (the title of her 2021 book), follows in Lasch's footsteps when giving the following description of what present-day neoliberal culture is doing to the key notions highlighted by Lasch, maturity and limits:

> Deregulating the morals of a nation to achieve their "person spec" would require a two-pronged approach: boost the uncaring part that wants to override limits and undermine the caring part that says no to that. […] Neoliberal culture infantilized and regressed people, encouraging them to

> be less mature and less responsible than they are capable of being; it seduced them into believing they could continue with business as usual. It used every trick in the psychological toolkit to persuade us that we need not change when only change will prevent systems' collapse. It drew society into a collective psychosis that people are now starting to wake up from.
>
> (Weintrobe 2021: 114)

I conclude this Introduction by giving a brief overview of the chapters that follow.

In Chapter 1, I take the influential book *A Perfect Moral Storm* by the American political philosopher Stephen Gardiner as my point of departure. Gardiner's book serves the purpose of introducing the topic at hand in its various dimensions, ranging from discounting (i.e., viewing and treating the present as more important and "valuable" than the future, hence future generations as less so than the presently living), to the role played by institutions and governments in "dealing with" climate change, to the significance of moral corruption, understood as the plethora of "good reasons" given for *not* taking the sort of action that the science considers necessary on the part of the most powerful players. Besides opening up the field of inquiry, Gardiner's book is discussed because it remains representative of the brand of political theory that has been hegemonic for the last 50 years: the political liberalism originally set out in John Rawls' *A Theory of Justice*. Due to its enduring influence, the shortcomings of this position when the issue faced is multidimensional environmental crisis will be a major topic. Indeed, I shall argue that a "nature deficit" – analytically as well as substantially, ethically as well as ideologically – is common to positions otherwise at loggerheads in political philosophy, be it in the form of Rawls and Gardiner on the one hand, and Michel Foucault and Judith Butler on the other. (I elaborate on this claim in Chapter 3.)

Chapter 2 is devoted to psychological approaches to the socially organized and culturally sustained forms of denial that I have outlined above. A key topic is the importance of shifting the emphasis from the "cognitive dissonance" exhibited in individual agents involved in a contradiction between what they know, or say, and what they do, to the emotional strain, typically lacking a proper language and so largely underarticulated, that individuals who do care, or used to care, about environmental crisis are suffering due to specific cultural patterns that "kill the conversation", making for feelings of powerlessness, apathy, and withdrawal. The empirically rich studies carried out by the psychoanalytically minded Australian psychologist Renee Lertzman and by the American (not Norwegian, although the name suggests it!) sociologist Kari Marie Norgaard, based on field work carried out in Michigan and in a Norwegian town, respectively, will occupy most of the chapter, allowing for in-depth case studies of "Sally" and of "Eirik". I contrast their contributions with that of the Norwegian psychologist and economist Per Espen Stoknes, allowing me to point out differences as well as similarities between how denial is socially organized in an American city and in a Norwegian rural town. Norway's position as a major global exporter of oil will be of particular relevance here.

Chapter 3 examines what it means to live in a neoliberal culture, especially with regard to the demand to build "resilience" that has become so prominent in recent years. I show the implications of the individualistic fashion this demand currently takes, highlighting the reversal in the relationship between institutions and individual(s) mentioned above: how, that is, individual agents are expected to shoulder the task and the burden of "adaptation" to ongoing climate change and environmental degradation in a society where the institutions, owing to neoliberal reforms, have increasingly abandoned their mission to provide security for their citizens, especially the needy and most vulnerable. The "cult of personal responsibility" plays a major part in making this reversal happen. A key finding in this chapter is that there is a rather surprising agreement between promoters of neoliberalism and its leftist critics: albeit for different reasons, both subscribe to a deep distrust of institutions in general, of government(s) and the state in particular, when it comes to their duties towards citizens. The result of this consensus across the ideological aisle is that institutions continue to decline, in an atmosphere of "diminished expectations" (Lasch), and with heavy costs to be paid by the exposed individuals, that is, by people and communities adversely affected by accelerating climate change and environmental deterioration.

Chapter 4 undertakes a shift, at long last, from the predominantly problem-oriented discussions of the earlier chapters to issues pertaining to the potential for resistance. *How to Blow Up a Pipeline*, the controversial book – manifesto – by the Swedish Marxist scholar and activist Andreas Malm is at the heart of this debate. Drawing on my own experience as an expert witness in two court cases over climate activism and with various forms of civil disobedience, I take readers from the arguments made in the scholarly books to the public sphere and the streets of Oslo, or London, where social movements building resistance, as opposed to individualized resilience, are actively confronting the powers that be. I show how such resistance is increasingly met with measures amounting to a thoroughgoing "criminalization of protest", whereby the activists are taken to court, beaten up by the police or silenced in less violent, more subtle ways, some winding up in jail, others in the therapy room or in the psychiatric ward, still others simply giving up, giving in, exhausted and disillusioned with what has become of "liberal democracies" facing the existential jeopardy of environmental emergency. Nevertheless, I argue against abandoning the expectation that institutions and government(s) can and should be part of the solution, not simply part of the problem. I highlight the role of whistle-blowers, focusing on the cases of Rachel Carson and Gretha Thunberg, with a keen eye for the difference increasingly made – so it seems – by gender, attested to by the aggression many white, 50-plus men express when met with demands to change their fossil lifestyle.

I conclude the book with a look at the speech made by Czech president Vaclav Havel in 1990. It serves as a sorely needed reminder not only of what a political speech can be, when at its best, but also of what political leadership can, and should, aspire to. Contrary to the statement made by one of the men Hochschild talked with in Kentucky, that "the first thing a political leader appeals to is fear, then to grief, then pride and shame" (2024: 25), the dissident-made-into-president Havel exhibits a quite different set of values when addressing his audience, treating them as adults and citizens: the sequence from truth-telling – his start, startling in view of the present-day Trumpian climate of post-fact and post-truth politics, goes like this: "I assume you did not propose me for this office so that I, too, would lie to you" – over admission of responsibility for mistakes made, to the necessity of getting the difficult feelings of shame and guilt that such responsibility gives rise to out in the open, for all to see, so that reparation and restoration of what has been destroyed – from trust in each other to a flourishing environment – can become possible, in a shared effort devoted to the common good. It says a lot about the "diminishing expectations" haunting present-day society that such leadership now appears so distant and beyond reach. It need not be.

Bibliography

Beckert, Jens, 2024. *Verkaufte Zukunft. Warum der Kampf gegen den Klimawandel zu scheitern droht*. Berlin: Suhrkamp.

Carrington, Damian, 2024. "We live in an age of fools", in *The Guardian Weekly*, 17 May.

Carson, Rachel, 1962. *Silent Spring*. Boston: Houghton Mifflin.

Freud, Sigmund, 1985 [1915]. "Repression", in *On Metapsychology*, 145–158. Harmondsworth: Penguin.

Gardiner, Stephen M., 2011. *A Perfect Moral Storm. The Ethical Tragedy of Climate Change*. Oxford: Oxford University Press.

Goffman, Erving, 1959. *The Presentation of Life in Everyday Life*. Harmondsworth: Penguin.

Goffman, Erving, 1961. *Asylums. Essays on the Social Situation of Mental Patients and Other Inmates*. Harmondsworth: Penguin.

Hamilton, Clive, 2013. "*What history can teach us about climate change denial*", in Sally Weintrobe, ed., 16–32.

Hamilton, Clive, 2014. *Earth Masters. The Dawn of the Age of Climate Engineering*. New Haven: Yale University Press.

Harvey, David, 2005. *Neoliberalism. A Brief History*. Oxford: Oxford University Press.

Hochschild, Arlie Russell, 2016. *Strangers in Their Own Land. Anger and Mourning on the American Right*. New York: The New Press.
Hochschild, Arlie Russell, 2024. *Stolen Pride. Loss, Shame, and the Rise of the Right*. New York: The Free Press.
Horkheimer, Max, and Theodor W. Adorno, 1979 [1947]. *Dialectic of Enlightenment*. London: Verso.
Hylland Eriksen, Thomas, 2024. *Det umistelige. Fra global ensretting til et nytt mangfold*. Oslo: Aschehoug.
Lasch, Christopher, 1990 [1979]. *The Culture of Narcissism. American Life in an Age of Diminishing Expectations*. New York: Norton.'
Lasch, Christopher, 1984. *The Minimal Self. Psychic Survival in Troubled Times*. New York: Norton.
Lertzman, Renee, 2015. *Environmental Melancholia: Psychoanalytic Dimensions of Engagement*. New York: Routledge.
Malm, Andreas, 2021. *How to Blow Up a Pipeline. Learning to Fight in a World on Fire*. New York: Verso.
Milman, Oliver, 2024. "Heroes or villains?", in *The Guardian Weekly* 2 August.
Monbiot, George, and Peter Hutchison, 2024. *The Invisible Doctrine. The Secrete History of Neoliberalism*. London: Allen Lane.
Norgaard, Kari Marie, 2011. *Living in Denial. Climate Change, Emotions, and Everyday Life*. Cambridge, MA.: MIT Press.
Rawls, John, 1971. *A Theory of Justice*. Oxford: Oxford University Press.
Stoknes, Per Espen, 2015. *What We Think About When We Try Not to Think About Global Warming*. White River Junction, Vermont: Chelsea Green Publishing.
Weintrobe, Sally, ed., 2013. *Engaging with Climate Change*. London: Routledge.
Weintrobe, Sally, 2021. *Psychological Roots of the Climate Crisis. Neoliberal Exceptionalism and the Culture of Uncare*. New York: Bloomsbury.

1 No way out

Varieties of a perfect moral storm

1.1 Cultivating complacency

Fifteen years after its publication, *A Perfect Moral Storm: The Ethical Tragedy of Climate Change* by the American philosopher Stephen M. Gardiner is still widely regarded as a landmark, offering an impressive interdisciplinary synthesis of the kind that its subject matter calls for. Clearly, if we search for answers to the question "What kind of a problem is climate change?" and "Why does it persist?", contributions from disciplines such as economics, political science, and psychology are required, providing the ethical assessment that is Gardiner's primary focus as a philosopher with a broad empirical basis.

For my purposes in what follows, what Gardiner chooses *not* to address is no less important than what he does address, often brilliantly at that. Of course, no book and no author can hope to exhaust all the aspects of such a multi-layered phenomenon as climate change: all the causes bringing it about and all the actors involved. To understand why the issues Gardiner either explicitly omits from his approach, or implicitly neglects, matter so much, it is necessary to take a close look at how he presents his book's agenda.

Gardiner takes a perfect moral storm to involve "the unusual intersection of a number of serious, and mutually reinforcing, problems, which creates an unusual and perhaps unprecedented challenge" ((2011: 7). Three different storms demand particular attention. The first storm is *global*, its key feature being that "the world's most affluent nations, and especially the rich within those nations, have considerable power to shape what is done, and to do so in ways which favour their own concerns" (7). The second storm is *intergenerational*: the current generation enjoys asymmetric power over the prospects of future generations. Each of these two storms involves a serious asymmetry of power, enabling some agents to take undue advantage of others. The "temptations" to which we are thus subject make us vulnerable to a certain kind of corruption, a fact that Gardiner will spend large portions of

DOI: 10.4324/9781003660743-2

his book attending to, both from a virtue ethical and from a psychological point of view. The third storm is *theoretical*. According to Gardiner, existing general theories, including intergenerational ethics and international justice, cannot guide us, being "extremely underdeveloped in many of the relevant areas", thus rendering us more vulnerable to the first two storms (7).

While each of the three storms undermines ethical action, what makes the storm characteristic of our present situation a "perfect" one is that taken together they are mutually reinforcing. The result is that instead of solving the problem, it goes from bad to worse. This is even truer today than it was when Gardiner wrote his book 15 years ago. To him, being caught up in a perfect moral storm means to be tempted to "distort our moral sensibilities in order to facilitate the exploitation of our global and generational position". Gardiner calls this "the problem of moral corruption" (8).

Since Gardiner from the start puts such emphasis on the ways in which "we" are vulnerable to moral corruption, the question becomes: who make up this "we"? Surprisingly, Gardiner does not raise the question, let alone provide an answer. It seems that once he formulates the issue as one of the psychology at work in moral corruption, a universal and as it were generic human psychological makeup is what is involved, one allowing for the same description in all human agents, regardless of whatever differences and inequalities that obtain between them, be it in terms of class, gender, age, ethnicity, and so on. The tacit assumption is that humans are – broadly speaking, which suffices here – all the same, prone to the same mechanisms of psychological defence in the face of danger or discomfort, and vulnerable to the same kinds of moral corruption. That this is a realistic assumption is held to be demonstrated in the tendency to exploit the possibility to take "undue" advantage of others and so to help reproduce existing asymmetry of power as long as doing so is seen as advantageous to oneself and one's own group.

Even though Gardiner employs an "all across the board" psychology when identifying the basic human motivations and dispositions operative in corruption, he does say something that is relevant to the matter. "My objective", he points out, "is not to vilify any particular individuals, groups, classes, or nations. Indeed, the point is not really to apportion blame at all, but rather to help us understand our own predicament" (6). But of course, as Gardiner knows only too well, the responsibility for what he routinely refers to as "our own predicament", suggesting a roughly equal part in it by all of us, is not at all equally shared, but instead characterized by enormous differences – economically and in terms of power in all its dimensions, material, and cultural.

Since Gardiner's one-description-fits-all way of portraying "our" situation is everything but true to the empirical realities in a world of obscene inequalities of power as to how concrete people are implicit in causing the crisis as well as in the degrees to which they are adversely affected by it, it does seem strange that he imputes this kind of realism to his proffered general description. He surely knows that it distorts rather than clarifies the situation. Why then does he choose this approach?

I see several possible answers. Gardiner may think that identifying the people whose actions and lifestyles carry most responsibility for climate change, and in that sense blaming them, is useless – useless because it will not help solve the problem by bringing about the necessary changes; or useless in the kind of book he is writing, being an academic work as distinct from, say, a political pamphlet. (Herve Kemp's book *How the Rich are Destroying the Earth* comes to mind as an aptly titled example.) The latter reasoning attests to a "division of labour" where academic books need to be non-polemical, descriptive, and fact-oriented as opposed to overly political and normative, and thereby taking sides in controversial issues and risk being seen as partial by some and so failing to be taken seriously by all. To apportion blame, then, would mean to fail to live up to the well-established requirements of moral and political neutrality that are integral to doing science.

The first assumption – that blaming specific individuals will backfire rather than help bring about change, and should therefore be avoided – subscribes to a psychological theory I find highly problematic, for reasons I shall say more about later. The second assumption is more relevant at this juncture, since I take it to expose key premises in Gardiner's book that he fails to lay out to his readers so that they can see whether they agree or not.

To appreciate what I am getting at, consider this statement: "As a matter of public discourse, the geopolitical disaster has been facilitated by the fact that the current generation in the developed countries has spent much of the last two decades conveniently distracted and confused about the problem" (9). This may seem evidently true, and let's disregard for now the lack of sociological accuracy involved in describing all individuals who make up the same generation, in the same kind of countries, in such a general and simplified manner. Key to the present situation, then, is that "few seem interested in really dealing with the problem despite its catastrophic potential". Admittedly, Gardiner continues, "this seems difficult to explain away in any normal way". So how to explain it? Well, unfortunately, he informs us, "it makes perfect sense" – it simply reflects the fact that we are in a perfect moral storm. This being so, "the temptation to pass the buck to the future, the poor, and nature is very strong. So, the incentive to disengage is high" (9).

Since what Gardiner says here may strike most readers as compelling, having perhaps encountered the same explanation and the largely tacit premises it rests upon countless times before, it is imperative that we pay it close attention. First, there is the paradox that a problem of "catastrophic potential" attracts the interest of only a few. Indeed, if the problem of climate change is a catastrophic one in the sense intended by Gardiner, so that it affects all people presently living, it seems strange that so many people take no interest in addressing it. Isn't it a matter of simple self-interest that they do so? Another way of putting the paradox is the frequently rehearsed one of a contradiction between what "we" know – as a result of scientific knowledge widely shared in society – and what we are prepared to do as a result of that knowledge, whereby the appropriate response would involve changing our ways, insofar as going on as before is seen as unacceptable, or even as downright dangerous.

How, then, can this paradox be explained? Gardiner's answer takes the form of removing the paradox: what starts out as cause for outcry, as a highly unexpected response – namely, that most people stay indifferent in the face of catastrophe – is said to "make perfect sense" once we understand that it is matter of a perfect moral storm, showing us how living in that storm, being part of it, makes us immune to it, as it were; immune to its reality as danger and so as a situation of urgency, calling for rapid and thoroughgoing change. As we saw, this assumedly widespread non response, the one that appeared so outrageous and contra-intuitive when first referred to, is what is to be expected, is what must be seen to make sense, since "the temptation to pass the buck on to the future, the poor, and nature is very strong" (9).

But why, exactly, is that temptation so strong? Why is the incentive to disengage so high, meaning higher than incentives to engage? Is it equally so for different individuals, belonging to different classes, different regions, and so on?

My point is that Gardiner commits the fallacy of naturalization: he naturalizes the non-response he sets out to explain to his readers, presenting it as precisely the normal or natural one, generalizing it such that it applies to all actors involved. On closer scrutiny, instead of answering the original question, the explanation offered only produces new ones: are human beings born with the disposition to "pass the buck on to the future, the poor, and nature"? Or to take "undue advantage of others"? Does not the very term "undue advantage" reveal some specific, meaning non-universal and so historical and cultural context, one where the distinction between fair and undue advantage is drawn to the individuals by way of example and socialization, such that how the distinction is drawn, and sought practiced, differs from one culture, or one societal context, to the next? If this is so, what justifies Gardiner in making the sweeping generalizations about human motives and incentives he relies upon in answering his own question?

This may certainly be regarded as a cheap point. And it can safely be said that Gardiner is aware that, say, cultural differences are real and that they detract from the validity of his explanation taken as one meant to apply to all people involved. Even if admitting as much, he may point out that his kind of inquiry cannot do without a certain set of general assumptions about human psychology.

Before I say more about this issue, we should note how Gardiner describes his own approach. Observing that the perfect moral storm poses "a very serious and deep ethical challenge", "mine is not", he assures us, "a council of despair" (10). What does he mean by saying this? "I am optimistic", he goes on, "that most of us have such concerns, and take them seriously" (10) – referring to the very concerns about what we are doing to the poor, the future, and nature that he one page earlier claimed that we are sorely lacking, apparently for reasons to do with our shared human nature. What are his reasons for suddenly voicing optimism, given the pessimism – based on what is portrayed as hard-nosed realism – about how we humans typically behave, pessimism in the sense that the reasons for our failure with regard to climate change are presented as psychologically deep-seated and universal?

Perhaps the optimism Gardiner suddenly articulates is mere lip-service; unfounded in his prior arguments and by the sorry state of the world, yet a gesture he feels obliged to offer, as if anything less than optimism would put his readers off, besides being regarded as less-than-helpful by politicians and activists alike. In my view, however, this kind of optimism – both as something taken to be expected, even demanded by the audience, and as something loyally, almost ritually subscribed to by authors – is part of the problem: it reproduces and reinforces the impasse in which Western societies find themselves, adding no sense of ethical urgency and political clout to dealing with the danger at hand.

Not that Gardiner pretends that we are on a safe course – far from it. To be sure, we are in the midst of a storm, and so far we don't cope at all. This being so, he balances his professed optimism with several warnings. For a start, we need to recognize that "the biggest obstacles to effective action may be our own complacency" (10). That may well be true, but note that – again – complacency locates the problem in a psychological, or if you will virtue ethical, register: as individuals, we should beware of complacency; it undermines our prudence in ethical judgment and our sense of responsibility to set things right. But how does this complacency arise? Why is it so widespread? Better still, who precisely has an interest in fostering it? Can we point at specific individual agents, or should we rather search for drivers on a higher level, that of societal structures, dominated by powerful corporations, such as fossil companies? Can it be that some of the most powerful agents in our society are capitalist corporations; agents more keen to *cultivate* complacency among people at large – taken as customers rather than citizens – than to commit (other than rhetorically) to solving the climate crisis, involving as it must that consumption in many areas needs to be reduced? Indeed, what about *conflict* – say, between different agents, many of them collective rather than made up of discreet individuals? What about the conflict between the goals – more precisely, systemic imperatives – of corporations (profit and growth) and the goal of a sustainable planet, dependent on sustainable levels of consumption, and so in stark contrast to proliferating trends of overconsumption? In short, what about the importance of conflicts of an altogether different kind than conflict within the individual – as in cognitive dissonance, as in contradicting oneself – and conflict between two or more individuals, or groups?

Having stated that complacency is the biggest obstacle, Gardiner advances a view of the challenge ahead inspired by virtue ethics. "The primary reasons for acting on climate change", he contends, "are not that doing so will be good (or at least not bad) *for us*; they are deeper and more morally serious than that" (11). Ethically speaking, then, what we cannot and should not do is "to engage in wilfull self-deception and moral corruption when the lives of future generations, the world's poor, and even the basic fabric of life on the planet is at stake" (11). Climate change should be a wake-up call, prompting us to "demand more of our institutions, our leaders, and ourselves" (11).

Before assessing whether this virtue ethical approach with its emphasis on the moral maturity of the individual agent is up to the task assigned to it, we need to look at the greater framework laid out in Gardiner's book. The mediation between the micro level of individual agency and the macro one of global structures will prove crucial to the plausibility of Gardiner's approach.

1.2 The intergenerational storm

What makes the storm Gardiner talks about *global* is the fact that it takes place within a historically specific spatial and temporal context. In a spatial perspective, what characterizes the storm is the dispersion of causes and effects, the fragmentation of agency, and prevalent institutional inadequacy. In keeping with Hans Jonas' well-known argument in his *The Imperative of Responsibility* (1984 [1979]), the consequences of the actions carried out by agents in modern societies are no longer limited in time and space to a context of proximity; instead consequences travel beyond the here-and-now, the face-to-face shared by individual actors. Thanks to ever more efficient and long-reaching technologies, the impact of actions transcends, eludes, and sometimes contradicts, the intentions of their individual authors, playing themselves out in spatio-temporal settings over which the actor has little control and is at a loss to anticipate. The disconnect thus made for between agent and intention on the one hand, and factual consequences for others, at other times and places, on the other, is typically experienced as a matter of "alienation" on the part of the individual, not being able to recognize herself in her actions and the world they help create. Add to this the fact that what truly makes a difference in the current situation of a storm, is the impact of *cumulative* consequences of actions stemming from *collective* agents – structures – and not only individual ones. I shall come back to the limits this imposes on a theory that remains committed to the proximity-based framework from which virtue ethics originates.

Despite this tension between the close at hand and the distant, it is among the merits of Gardiner's book that he highlights what he calls "the intergenerational storm". His reasoning here starts from the physical fact that carbon dioxide is a long-lived greenhouse gas. This has three important implications. First, climate change is a *resilient* phenomenon, rendering the upwards trend in atmospheric concentration virtually irreversible. Second, impacts of climate change are *seriously backloaded*, being the result of emissions from some time in the past, having often taken place somewhere else than where they currently, or eventually, exert their impact. Third, and accordingly, we have to do with a *substantially deferred* phenomenon, such that the "full, cumulative effects of our current emissions will not be realized for some time in the future" (34).

Accompanying these three features of climate change is a twofold fragmentation of agency: a spatial one and a temporal one. Whereas spatially fragmented agents – say, in different countries – may become unified and so able to act as a single agent, temporally fragmented agents cannot actually become unified; at best, Gardiner observes, they may "act *as if* they were a single agent" (34). This being so, "temporal fragmentation may be more intractable than spatial fragmentation" (35). We face "a new kind of collective action problem that not only adds to the original storm, but is also more difficult to resolve" (35). Gardiner designates this problem as that of "inter-generational buck-passing".

Why is temporal fragmentation more intractable than the spatial one? Gardiner answers by way of pointing out that different generations have different interests when it comes to how they assess the urgency of reducing emissions. Whereas "the benefits of carbon dioxide emissions are felt primarily by the present generation, the costs are substantially deferred to future generations" (35). In this situation, the current generation "may 'live large' and pass the bill on to the future" (35). As long as generation-relative concerns continue to prevail, the trend of overconsumption will persist. What Gardiner refers to as the "tyranny of the contemporary" will go on unabated, with meagre prospects of being halted.

Convincing as these points may seem, their validity depends on some assumptions that are far from unproblematic. By way of definition, Gardiner conceives the intergenerational storm in such a way that the parties do not coexist; hence "the afflicted are in principle unable to directly influence the behaviour of their predecessors" (37). This makes for an important contrast with the well-known "prisoner's dilemma" and "tragedy of the commons", both discussed by Gardiner as part of his standard repertoire as an American political scientist. Since in both these cases the parties coexist, whereas in the intergenerational one they do not, the latter cannot appeal to mutually beneficent interaction, which is to say that reciprocity is sorely lacking, and so the possibilities of give-and-take, implying a break on pure self-interest, that might appear attractive to both parties. Again, Gardiner portrays the resultant situation in categorical terms: "Between present and future generations there is no interaction at all (by definition), and so no mutual benefit (there is no way for future generations to benefit present generations)" (37).

In my view, for all its appeal to common sense, this statement is simply not true. For a start, Gardiner's definition of the intergenerational problem is too categorical and rigid. Of course, there is such a thing as generations of the past (all dead), of the present (people now living), and of the future (the not-yet-born). But there is no reason to restrict, by definitional fiat, the challenges posed by climate change to different generations, to this set-up, one where generations neither coexist nor interact with each other, but merely follow in linear succession, the one "taking over" when its predecessors are dead, dominating the course of things completely with regard to future generations understood as those not-yet-born. The "one generation at a time" model employed by Gardiner fails to convince both empirically and theoretically, rendering his approach to the temporal dimension much less helpful than it sets out to be.

What, then, would a more convincing model look like? At present, as far as the impacts of ongoing climate change are concerned, three generations are affected, in diverse ways at that: children, their parents, and their grandparents (see Vetlesen and Willig 2018). They coexist in every sense of the word – not only spatio-temporally but also psychologically, emotionally, and ethically as well – and, again from diverse vantage-points, each of the three generations qua coexisting has ample opportunity to register, and to respond to, to seek to influence, the other two.

The by far best known illustration of this point may well be the speeches made by Swedish climate activist Gretha Thunberg, who single-handedly, at the age of 15, started the greatest political movements by young people ever seen, following her lone protest in front of the parliament in Stockholm in August 2018. Having risen to fame in a matter of months, Thunberg got the chance to speak truth to power in one country after the other, reaching a peak with her speech in the UN General Assembly in September 2019. Her message was the following one: "Your generation is failing us. But the young people are starting to understand your betrayal. The eyes of all future generations are upon you. And if you choose to fail us I say we will never forgive you" (Thunberg 2019. 99).

In the company of Gardiner's sophisticated book on climate change, it may seem banal to give this example as proof of the limits of his approach to the intergenerational storm. Be that as it may. Of course, more needs to be said to assess the importance of the kind of intervention made by Thunberg, first in her splendid isolation, later as a leader of tens of millions of young people worldwide; I shall return to this topic when discussing resistance in Chapter 4. The point I am making here is not merely empirical;

it is theoretical as well. It is that Gardiner commits the fallacy of conceiving generations not only in two rigid a manner – one at a time – but also that he imposes a sort of *homogeneity* – of perception, of interests, of priorities – upon each of the three (non-coexisting) generations his typology allows for that is simply not true to the facts. It is a theoretical and conceptual simplification that is not only as such unfortunate and unnecessary. More central to my concerns, it also distorts the complexities that play out *within* each generation, concealing the tensions and conflicts between the youngest and the old(est) members of presently living people (Thunberg), and that between peers (say, 40- and 50-years old) as well, as in situations where, for example, the importance of class differences may outweigh that of being of the same age.

There is another dimension to this that is more subtle. It is to do with Gardiner's book being written a good 15 years ago. Has something happened since then that makes his argument less convincing than it appeared at the time? I take it that the reply to this is rather simple: the crisis he describes has become more urgent. The crisis is more serious, our situation more dangerous, the obstacles to averting "the worst" – runaway climate change, crossing the nine tipping points identified by Johan Rockström et al. (2009) – evidently going even deeper, being more recalcitrant and persistent than Gardiner thought – and this even in the face of ever-mounting evidence of its realness, its dangers and risks.

Indeed, the crisis has gotten worse since Gardiner wrote his book, and there is no reason to document the well-recognized fact here. My intention is not to fault Gardiner for underestimating the mess we're in – he does sound seriously worried, at least as worried as his lip-service to optimism permits. What is important, rather, is this: the consequences of climate change, we presently see, are not so overwhelmingly part of the future – understood as a world where we presently living are dead, a world to be taken over by the "next" generation – as distinct from the world of the present that Gardiner presupposes in his model. In talking about the current generation as one – homogeneously – allowing itself to "live large" and pass the bill on to future ones, Gardiner glosses over the ways in which climate change took its toll and endangered the lives of many people "already" when he wrote his book in 2011. Although this applies to more people today, say in his United States, than at the time, the fact attests to an unfortunate tendency in Gardiner to paint the costs – in more than economic terms – of climate change as *mainly* costs-to-appear-in-the-future, hence as a problem that, given the lack of will of the presently living to deal with it, will overwhelmingly harm the lives of those who come after us.

Again, to the extent that there is some truth to this proposition, it is less so with each passing day. This has bearing on the often implicit psychological assumptions on Gardiner's part. To simplify somewhat, the constellation he paints between the present and the future is that between climate change appearing as "ain't seen much yet", and so as a relatively distant and only modestly experienced reality, to being – indeed – first-hand experienced, in a not-to-be-denied-as-real manner, as taking place here-and-now, hence a change that is utterly concrete, and perhaps so averse and dangerous in its impact – here where we live now – that moving to someplace else, hopefully safer, less vulnerable to the dynamics unleashed (the heat rendering agriculture impossible, the water disappearing, the animals, including fish, dying, and etcetera) becomes imperative. In other words, Gardiner's way of drawing the distinction between the present and the future – reserving the really serious consequences for the future understood as people as yet unborn – gives a mistaken picture. It needs to be replaced by a more dynamic, erratic, and volatile outlook, where the present is impacted – disrupted, endangered – by changes that Gardiner for the most part locates in the future. Whereas the impacts of climate change are continually *expanding* in time and space, the temporal frame – window – where action must take place is dramatically *shrinking*, into a more and more compressed present at that. In this present the ethical demands on the current generation of adults do not reside in the future understood as the not-yet-born generations that will succeed us; rather, the demands are articulated here and now by our own children – a generation, that is, for whom the question of whether they will want to have children is becoming an existential issue in a sense not experienced by previous generations.

1.3 Rewarding the short term

In his main theoretical discussion of the intergenerational storm, Gardiner singles out "asymmetric independence of interests" as a particularly crucial feature. He defines the asymmetry as follows: "Though later groups have a substantial amount to gain from earlier groups, the reverse is not true: the interests of earlier groups are independent of the groups which succeed them." Therefore, "earlier groups have nothing to gain or lose from the activities of later groups (though later groups have a substantial amount to gain from earlier groups)". The upshot of this feature, then, is that it "rules out the possibility of intertemporal exchange for mutual advantage" (167).

Philosophically, the problem with this presumably purely academic way of defining "interests" is that it buys into the very concept that interests are something possessed by the individual in question: by default, interests are self-oriented, those of the self for itself, not for others. Gardiner concedes as much when he says that "the assumption is that each group has preferences that are exclusively generation-relative: concerned with events that happen during the timeframe of its own existence" (167). In this statement, the self in question is that of a group – my generation – rather than the individual as such; whereby interests have been translated into preferences, with no clear demarcation made between the two. Preceding this statement is Gardiner's observation that "the benefits from abstaining [from energy consumption in the present] are likely to be small, uncertain, intangible, and deferred", thus making for a stark contrast with the benefits of instant gratification. As we saw above, for Gardiner the contrast between the present and the future is that between the concrete and the abstract, between the certainties allowed us in the here-and-now and the uncertainties of a more or less distant future. The overriding assumption is that the present will "naturally", quasi inevitably prevail over the future as far as our priorities, tied to our interests and articulated in our preferences, are concerned.

To see why this assumption is problematic, we turn to his discussion of "personal attachment" as a possible objection to his model – that is to say, the idea that "attachment can ground strong concern for later generations with power to override (or modify) self-interest or generation-relative aims" (172). Gardiner comments that holding personal attachment to be important means to presume that "attachment occurs only on contact", adding that "contact might be a long time coming, and so too late" (173). Can I feel personal attachment to someone – or even something – I presently have no contact with, and in that sense do not interact with? Well, I feel closely attached to a number of people who are dead, whereby lack of opportunity to interact is a quality sorely missed, yet does not as such detract from my emotional, memory-based attachment to them; rather it will remain with me as long as I live. Now of course, in Gardiner's argument the relation to the past is not the central one; the possibility of a relation to the future is, and this is what Gardiner in effect rules out. But again, his conception of temporality is too rigid and too mechanical. My attachment to my children, and to grandchildren who are not yet, but someday may be born, be it in my lifetime or not, is certainly something I feel in the present, with its addressee coexisting with me in the first case, not so in the other. What holds for both cases, though, is that my attachment is *future-oriented*, projected into a future that, even though I may not be able to be part of it and help shape it, I hope will turn out for the best for those to whom I presently am attached. Echoing an insight from Heidegger's classic analysis of temporality in *Being and Time* (1927), the future-oriented – "thrown", projected – dimension of my present attachments and concerns is not separate from it, or an addition to it that may or may not obtain, but part and parcel of it: while always bound to a context here-and-now, what drives my attachment is the wish for and investment in a future where its addressees will thrive; hopefully, a future that, contrary to Gardiner's assumption of a "natural" favouring of the present over the future. will prove *better* – more allowing for flourishing, more safe and secure – than the present of which I am now part. This being so, the limit to the temporality of attachment that Gardiner presents as inseparable from it, as part of its nature, is nothing of the sort; it distorts and misses out on its intrinsic temporal structure as future-oriented not present-fixated.

Add to this Heidegger-inspired criticism of Gardiner's position an empirical one. Consider the following statement:

> Merely to assert that reciprocity and personal attachment are possible does nothing to show that they are actual, that they are present now, or (most importantly) that they are strong enough to overwhelm the competing influences of self-interested or generation-relative concerns.
>
> (173)

This raises the question: what is the origin and nature of "the competing influences" exerted by the latter? This is a critical question because Gardiner defines interests and generation relative concerns the way he does: as prioritizing the self over others, and the generation of which one is part over others, particularly future ones understood as the not-yet-born. But are people generically, intrinsically, self-interested in this manner? Given that there are, as Gardiner himself implies, "competing influences" in play, how strong is the assumed "natural" pull in the direction of acting on self-interest in Gardiner's postulated sense, and how strong is the pull to counter it and act differently? Does the latter pull reside within the individual – as in being conflicted with respect to self-interest versus other-directedness – or does it emanate from "outside", say, from cultural influences and incentives?

Posed like this, the question is admittedly naïve: of course the answer is that both obtain and help determine the priorities borne out in actual behaviour. Thus the more interesting question concerns the relative strength of the different pulls at work. To discuss this is a less abstract manner, I shall exemplify what I am getting at by focusing on the phenomenon of short-sightedness.

Conducive to my purposes, at this point in his discussion Gardiner addresses the role played by institutions. "Governments", he writes,

> are often focused on their impacts over limited terms of office, particularly as they affect their ability to win the next election; corporations are often focused on the dividends likely to be produced in the immediate years ahead, not the further implications of their actions.
>
> (173f)

With regard to central institutions such as governments and corporations, then, "the presence of extremely limited time-horizons is obvious" (173). Why obvious? If anything, is not the core function of the institutions characteristic of modern society that they *compensate for*, and serve as *correctives to*, the limited time-horizon of individual actors who come and go while the institutions survive them and transcend their limits? Indeed, to suggest that major institutions are inherently short-sighted or prone to becoming ever more flies in the face of how the classic sociologists – Max Weber in particular – defined their very function and their legitimacy, not forgetting that another classic, Marx's analysis of capitalism, set out to show how the systemic imperatives of profit and growth exert a strong pull in the direction of prioritizing short time horizons over long ones. This tendency, however, is not part and parcel of institutions *per se*, but instead characteristic of how they operate in a specific historical and cultural context, one allowing – at least in principle – for thoroughgoing change.

I take this to show that Gardiner here, as so often in his book, naturalizes the phenomena he discusses. This is brought out when he seeks to explain the "obvious" fact that institutions in contemporary society – Gardiner has the United States in mind – are committed to "extremely limited time-horizons", pointing out that "such institutions are typically headed by elites, who are predominantly people in their mid-forties to mid-sixties" (and overwhelmingly male, one should add). Within these group, he tells us, "what is important is to make a good, highly visible, short-term impact in a given institution, as a way to move on. And this results in an incentive to ignore the long-term implications of policy" (174).

Make no mistake: I do not deny that in present-day society governments and corporations alike operate according to ever more limited time-horizons, or that their leaders are committed to this trend (see Wolin 2018; Hylland Eriksen 2016). The question is why – how to explain it? Why, that is, is the prevalent situation such that leaders are *rewarded* for prioritizing short-term over long-term perspectives and consequences? Is it a matter of their choosing to be thus rewarded, or is a matter of demands not of their own choice but demands – imperatives – to which they need to adjust, regardless of what they think about them qua individuals, from the moment they enter the institution – such that if they, say, favour prioritizing the long-term over the short-term, they will either be fired or not get the job in the first place? And if this is indeed the case, do we not need to shift the perspective from the individual agent to the larger structures – with its inbuilt imperatives – that he or she is part of?

In my view, Gardiner is guilty of a conflation of causes and consequences. Rather than holding, as he does, that the limited time horizon on display in the actions of the individuals in question "result in an incentive to ignore the long-term implications of policy", it is the other way around: the latter incentives *await* the individuals who come to occupy these positions; they predate the trajectories of leaders who come and go, who succeed and fail, and they do so because they have become intrinsic to the structures entered into – intrinsic not in the sense of being – timeless, universal – features of these institutions *per se*, but in the more contingent, historically and culturally specific sense of having, especially in the last 40 years, since the onset of neoliberalism, become practically inseparable from them, so that opposing the incentive to let the short-term prevail, or otherwise failing to satisfy it, will be punished, the individuals involved replaced.

To appreciate some key aspects missing in Gardiner's very general portrait of the modern corporation, Joel Bakan's book *The Corporation* is useful. Having been granted the juridical status as a legally designated "person" dedicated to valorize its self-interest and to invalidate all sorts of moral concern that do not strictly bear on its internal dynamics and goals, most people, Bakan observes, "would find the corporation's "personality" abhorrent, even psychopathic, in a human being, yet curiously we accept it in society's most powerful institution" (2005: 28). Milton Friedman, the well-known economist who more strongly than any other academic influenced the neoliberal policy reforms introduced by Ronald Reagan and Margaret Thatcher (see Harvey 2005; Mirowski 2014), tells Bakan what he has conveyed to his wide readership in his best-selling books: "A corporation is the property of its stockholders." Hence, there is but one "social responsibility" for corporate executives: "they must make as much money as possible for their shareholders. This is a moral imperative. Executives who choose social and environmental goals over profits – who try to act morally – are in fact, immoral" (2005: 34). Though the latter is Balkan's wording, it accurately captures the essence of Milton's teaching. As Balkan explains, the "best interests of the corporation" principle, now a fixture in corporate laws in most countries,

> forbids any other motivation for the actions of corporate decision makers, whether to assist workers, improve the environment, or help consumers save money. [...] Corporate officers have no legal authority to pursue such goals as ends in themselves – only as means to serve the corporation's own interests, which generally means to maximize the wealth of its stakeholders.
>
> (2005: 37)

Bakan tells the story of John Browne to illustrate his point. In 1999, Sir John Browne gave a speech in his capacity as head of BP (British Petroleum), at the time the world's second largest oil company and the largest single supplier of oil and gas in the United States. Browne broke ranks with his CEO-colleagues in the oil industry by publicly acknowledging that greenhouse gas emissions cause global warming, going on to endorse the Kyoto Protocol on reducing greenhouse-gas emissions and to pull out of the oil-led Global Climate Coalition, which had spent millions of dollars lobbying against the protocol (see Oreskes and Conway 2010). Presenting himself as the world's first "green" oil titan, Browne received an award on 22 April 1999, that year's Earth Day. In his speech he said that "We have to help people transcend the harsh trade off which says you can have economic growth and pollution, or you can have a clean environment but no growth. That's an unacceptable trade off" (quoted in Bakan 2005: 40).

For a CEO in a major oil company, this was a big step to make, unprecedented at the time, although many will be quick to point out that, insofar as Browne subscribes to the idea of "decoupling" – that economic growth may continue without harming the environment (a win-win scenario) – it falls far short of the radical changes that need to be made. What matters most from Bakan's point of view and in light of Gardiner's analysis, however, is that Browne raises the question "Can business be about more than profits?", answering that it can.

But to suggest that corporations and those who run them can be genuinely committed to values other than profit flies in the face of the law as defined above. Bakan put the issue to Friedman, prompting the following response:

> If a CEO pursues environmental interests in such a way as to run the corporation less effectively for its stockholders, then I think he is being immoral. He's an employee of the stockholders, however elevated his position may appear to be. As such, he has a very strong moral responsibility to them.
>
> (quoted in Bakan 2005: 42)

To many people, at least outside the fossil industry, this will come across as a truncated version of what "moral responsibility" is about; or put more critically, as its perversion, to employ a term we shall see is central to Gardiner. As for Browne, he would soon come around to let everyone know that he is no radical, after all, adhering to the oil business's credo that is also, presently no less than in the past, its bottom line: Facing the prospect of non-fossil alternatives, "we have to compete to ensure that oil remains a fuel of choice" (quoted in Bakan 2005: 46). So it not only that BP has failed to live up to transforming itself into "Beyond Petroleum" – the buzzword concocted for its advertising campaigns, inspired by Browne – it has also failed to enact the possibility of becoming Beyond Profit, taking the company back to start, which is to say where they are today, with Browne long gone: simply British Petroleum.

Though fascinating, there is nothing new under the sun about the story of John Browne as told by Bakan. Readers of Naomi Klein's 1999 breakthrough book, *No Logo*, describes the dynamics of "the race to the bottom" at work in moving entire industries (cars, manufacturing) from advanced industrial states such as the United States and Germany, to the regions of the world – especially Asia – where the costs of production are kept at an absolute minimum, such that trade unions are forbidden, environmental laws, and regulations either ineffective or non-existent, child labour rampant, and so on (see Klein 1999; Bauman 1998; Harvey 2005; Urry 2014; Lessenich 2016). Companies headed by CEOs hoping to make a difference, ethically speaking, are forced out of the game; only the toughest survive. To point out that this is the way companies operate, highlighting the logic they are compelled to adapt to and to enact even more cynically than their competitors, is to recall the analysis of capitalism that goes all the way back to Marx. The structural logic, that of the primacy of the twin-imperatives profit and growth – profit as depending on growth continuing; growth made possible by profits raising – remains the same, no matter how much else has changed since Marx published volume 1 of *Capital* in 1867.

1.4 Discounting and its fallacies

Return now to the point where we left Gardiner's analysis: the nature and significance of "extremely limited time-horizons".

Drawing on Bakan's study of modern corporations, I have argued that the tendency rightly observed by Gardiner – that time-horizons become shorter in a situation of climate change where they need to be extended – is not to be understood as features of individual agents. Rather it must be recognized as an intrinsic feature of corporations under capitalism, exacerbated in the present era. In other words, it is matter of structure, of systemic imperatives part and parcel of a capitalist economy, rather than of the psychological dispositions of the individuals who occupy the top positions of a specific company such as BP.

As far as agency and issues of responsibility are concerned, this is an important distinction. In Gardiner's case, his privileging of individual agency over structural features is not a matter of personal fancy, but follows naturally from his favoured ethical theory, virtue ethics. There is more to this than the choice of ethical theory, however – that is to say, in choosing virtue ethics of the sort going back to Aristotle over deontological (duty-oriented) ethics inspired by Kant and over utilitarian-consequentialist ethics inspired by Bentham. The fact that all three great traditions of Western moral philosophy are conceived with the motives and actions of individual agents in mind, though with increasing attention being paid, since the onset of modernity, to the "total" consequences of accumulated actions over time and across regions (Hans Jonas), is important to understanding Gardiner's approach. Equally crucial to my purposes, however, is the way the methodological individualism common to this whole intellectual legacy is tacitly built into the sort of liberal political theory that Gardiner, to a certain extent, sets out to criticize, yet remains deeply committed to.

The implications of subscribing to methodological individualism will occupy us for the remainder of this book. In my view, they are key to understanding not only why climate change is not adequately addressed in our society – Gardiner's major topic – but also to how it is that this crisis in the relationship between modern societies and the natural world has come to unfold in the first place, and so far is only getting worse.

In what follows, I shall argue that the phenomenon of *discounting* attests to the role played by an individual-focused approach and the models it gives rise to, in political theory as well as in economics. It may come as a surprise that discounting should be relevant in this way, used as we are to encounter it, if at all, in highly technical terms in the journals of economists.

To see its relevance, we need first to ask: what is discounting in the context of climate change and responses to it? At its core, and to quote Gardiner, discounting is the tendency to "reduce the weight of future costs and benefits relative to current costs and benefits" (267). In short, it is to *put more weight on the present than on the future*, so that – other things being equal – the benefits and costs involved in today's state of affairs take precedence over the benefits and costs of tomorrow's, and tomorrow's take precedence over next year's. The more distant into the future the state of affairs to be considered, the less significance and value will be ascribed to it relative to the present or what is closer to it. Note that while discounting thus defined is a phenomenon of temporality – the distinction between what is near and what is distant being one of time – it also clearly has a spatial dimension, so that what occurs close in space takes precedence over what occurs far away.

I have deliberately phrased discounting in a way that says nothing about its subject: who is engaging in it. But isn't it obvious that this subject is, and can be no other than, individual human actors, so that discounting is something to which *we* are typically prone? Are we not talking about a characteristic trait of us humans, deep-seated at that, so as to help constitute our human nature? And while recognized as being a shared, all-too-human disposition, it is of course one allowing for – calling for – ways of being handled, possibly resisted, by individual agents, making this an ethical task, and a proof of ethical virtue.

Though fine as far it goes, this common-sensical notion soon becomes too simplistic when we turn to climate change. For obvious reasons, what requires attention here is the way society, or companies, are involved in discounting, with the typical result that actions needed to avert "the worst" of climate change are being deferred, postponed, downplayed – again and again. It is as if though admittedly "important", even "urgent", such actions are never important and urgent *enough*, at least not for *us*, not *now* – there will always be reasons to pass the buck to the future, to invoke Gardiner's way of putting it. And since this is so, attesting to a regrettable truth about how we humans are, the repertoire of psychological mechanisms at our disposal will be all-too-well rehearsed, involving denial and all sorts of rationalizations to see to it that our reluctance to act, to change our ways, not harm our self-respect in a serious way.

Discounting in the form discussed by political philosophers like Gardiner, and prominent economists like Nicholas Stern and William Nordhaus, requires us to go beyond the individual-centred approach, asking: how can a society, or an economy, tackle the phenomenon of discounting? Note that the parties affected by discounting are typically conceived as generations – people living now as distinct from people not-yet-born – like we saw in Gardiner. Present is divided from future in the categorical, static, and reciprocity-excluding manner I pointed to, and criticized, above, and the same division is applied to values and interests (see Spash 2006: 203).

The result yielded by the model with which the Yale economist William Nordhaus is associated is that "only limited abatement should occur in the next twenty to thirty years, since the costs of current reductions are too high in comparison to the benefits" (Gardiner 2011: 257). By contrast, the British economist Lord Nicholas Stern, main author of the famous 2006 Stern Report, states that "the benefits of strong and early action far outweigh the economic costs of not acting" (258). As it stands, then, the two prominent economists disagree: while the one holds that only very limited action on climate change is justified, the other holds that comprehensive action – now – is justified.

What does "justified" mean in this context? Justified for whom? And justified at what time (not to mention what places on Earth)? These are simple questions, allowing however – it soon turns out – for no simple or conclusive answers by the economists debating them. Their disagreement persists, much like that – it is tempting to add – between philosophers debating whether there is intrinsic value in non-human nature, or only in human beings.

Before considering a possible answer to the question, we need to note the extent to which arriving at different conclusions is a result of differing premises in the models employed. In technical terms, there is the discrepancy between Nordhaus's discount rate of 5, 5%, and Stern's 1, 4%, which is to say that Nordhaus posits a greater difference between the present and the future than does Stern, such that the costs in the present for actions that have a positive impact in the future is regarded as higher by Nordhaus than by Stern, that is, too high for us now to be "worth it", whereas Stern emphasises that non-action now creates ever-greater costs for the future, and that these merit priority in the present: we cannot, with open eyes and resources at our hand, be responsible for making the situation worse, or steadily worse, for those who come after us *simply because they come after us*. Put differently, future lives are not inferior in terms of value to present ones, and so future conditions – shaping those lives – cannot weigh less, or significantly less, than present ones.

It will not escape the reader that this way of putting it is not only free of technical parlance; it is undeniably ethical, smuggling in as it were a notion about what would be the right thing to do. Some may hold that this step into ethics, into judgments of value, of right and wrong, virtues and vices, is and must be no-go area for economists; that they should provide us with the facts, and then it's up to us – as citizens, as politicians, and so on – to exercise moral judgment about how we should act.

Yet, according to the British economist Clive Spash, a long-standing critic of the hegemonic paradigm of neoclassical economics, the implied division of labour between fact-oriented economics and value-based ethics fails – it fails both to obtain and to convince. The approach epitomised by Nordhaus, far from ethically neutral or agnostic, at an arm's length from presuppositions and notions of value, is "actually defending a very particular ethical position rather than being free of all [ethical] considerations" (Spash 2006: 203). As we saw, the very definition of discounting is based on the – unquestioned and tacit – assumption that human individuals - regardless of history and culture, of whatever impacts on their motivations and behaviours – "naturally" prefer benefits now rather than later, "while costs are regarded as better delayed as long as possible" (2006: 204). Spash concludes his critical discussion by stating that "people forming their present and short-term preferences as consumers and investors in markets simply ignore the consequences in the remote future" (2006: 215). This being so, "discounting makes acting to degrade the future environment highly attractive"; "many economists wilfully fail to confront the ethical implications of discounting" (2006: 215).

While I agree with the criticisms advanced by Spash, more needs to be said about the intimate relationship between conceptions of discounting and ethics, giving the lie to the credo that the two are, and must be, kept strictly apart. Given that there is – for the sake of argument – a disposition to act on short-time preferences, how does it originate, and what helps cultivate it? Spash implies that in our roles as consumers and investors we act in markets that, for all practical purposes, *favour* that we pursue short-term as opposed to long-term preferences, making our complicity in degrading the future highly attractive, as we saw. That this way of acting is being favoured in these roles can be put more strongly still: we are being rewarded, not punished, for sticking to the short-term perspective; systematically and structurally so. It is not a matter of personal choice or preference. As Naomi Klein (2014) and a host of others have documented, the time frame differentiating those who succeed – as investors, as companies – from those who fail is shrinking, in a situation where the consequences following from the triumph of the short-term – in production and consumption alike, with the resultant overexploitation of limited resources worldwide – are expanding and getting increasingly severe, often irreversibly so.

The question prompted by this state of affairs can be put in a very simple manner, with full ethical force: how is it possible that being part of a situation where the future is degraded appears as "highly attractive"? Do we as human beings have a kind of natural predilection for bringing about this state of affairs? Indeed, wouldn't the intuition be the very opposite: that what we wish to be part of is helping make the future *better* than the present, not worse? Do we really want a much-degraded world to be our legacy, compelling our children and grandchildren to cope with the mess?

Putting the question like this helps us identify a feature of the dominant notion of discounting that will occupy us for the remainder of this book: whereas this notion posits as a fact about human nature that as individuals we care less about future states of affairs than current, and do so even if this implies a much-degraded future, the resulting preference for such – foreseen, expected – degradation appears dubious, if not downright counter-intuitive when considered as a proposed truth about us. This is important given Gardiner's discussion of discounting as conceived by leading economists, where – despite the disagreement between, say, Nordhaus and Stern – the contributors take the validity of this proposition as their "realistic" point of departure, so that rejecting it would betray believing that humans are more altruistic, more caring about others, be it distant others in time and space, than they really are. The realism – empirical validity – is provided by referring to how people – or most of them, most of the time – behave, like it or not: the behaviours we observe in present-day society, in consumers and investors alike, are proof that the short-term prevails. This being so, there is no contradiction, no conflict or tension, between how people behave qua individual agents and how corporations operate within the structures they are part of; between society's micro and macro level. Rather the picture painted is that of a smooth and "natural" continuity and *fit* between the two: the short-term priority of the one is reflected and borne out in that found in the other.

I agree with critics like Spash who regard this position as ideological not scientific, normative not value neutral, despite its constant claims to the contrary. It is a position where the *modus operandi* of present-day institutions are naturalized because painted as universal givens to be taken for granted. But again, there is no necessity or inevitability to institutions – governments as well as corporations – being committed to the short-term over the long-term, as though this should follow unavoidably from the "truth" about human behaviour. Not modern institutions *per se* function like this, but present-day capitalist ones – especially companies – do, this being a matter of a historical circumstances that in principle are open to change. As any student of history and social anthropology knows, in many epochs and cultures – some of them still existing, though hardly flourishing in our times of economic globalization and cultural homogenization – the time horizon entertained and aspired to by individuals and borne out in their institutions is that of "seven generations to come" (see Kopenawa and Albert 2013; Hylland Eriksen 2024; Vetlesen 2019). If anything, short-sightedness of the sort at display in today's Western societies is a recent phenomenon, conspicuous in how it contrasts with most of human history.

1.5 Taking "favourable conditions" for granted

Given this critique of the concept of discounting as presented by Gardiner, where does he stand? Does he side with the hegemonic notion?

I find it difficult to tell. Gardiner mentions two assumptions that merit particular attention. The first is the assumption that "human systems have become substantially independent of natural systems", that is to say, that "human life is by now fairly independent of the natural world", an assumption at the core of Nordhaus's model and regarded as too complacent by Stern. Gardiner gives the following quote from the prominent philosopher Thomas Schelling, a champion of game theory and rational choice, to illustrate the kind of optimism that goes with assuming such independence:

> Except for a very low probability of a very bad result…[Americans] are probably going to outgrow any vulnerability we have to climate change. And in any case we'll be able to afford to buy food or import if it is necessary. You know, very little of the US economy is susceptible to climate.
> (quoted in Gardiner 2011: 263)

I shall have more to say about this assumption below. Suffice it to observe here that the idea that human societies are becoming increasingly independent of the natural world – that is, the idea associated with the so-called "decoupling" – is completely at odds with what the science and the empirical data show: the inseparability of human activities and the ecosystems within which they occur and which they impact on and are impacted by is a hallmark of the Anthropocene; the many ways in which economy and ecology become *more* intertwined, more closely interconnected than at any earlier time in history. Contrary to Nordhaus's and Schelling's pronouncements, vulnerabilities of a number of sorts are rapidly proliferating as a result, increasingly making themselves felt in the present, not only in the more or less distant future (see Dalby 2009).

The second assumption singled out by Gardiner is that "environmental goods are infinitely substitutable for other goods"; hence, "improved productivity can always compensate for any other losses" (263). In this way, more consumption can always make up for environmental degradation, implying that "suitably bigger economies (and domes) are definitely available to us, and can be brought to bear very quickly" (263). Though nowhere mentioned by Gardiner, this recalls the pioneering Rumanian economist Nicholas Georgescu-Roegen's critique of what he called "the fallacy of endless substitution", attributed equally to the neoclassical school and to the Marxist one: "Nothing", writes Georgescu-Roegen, "could be further from the truth than the notion that the economic process is an isolated, circular affair". Rather, "the economic process is solidly anchored to a material base which is subject to definite constraints" (1970: 59). More clearly for all to see today than when Georgescu-Roegen made this point over 50 years ago, the sources of value – designated and treated as "resources" – that are undermined by current patterns of overconsumption and overexploitation of the natural world, taking the form of depletion and extinction, cannot, once their undermining is set in motion on the scale presently unfolding, ever be fully restored.

I will add to this a third assumption that fails to get the attention it deserves in Gardiner's book – namely, that all outcomes can be expressed in the same units, usually money, insofar as they are regarded as commensurable. Less abstractly put, consequences – say, of present-day lifestyles for future generations – are seen as "measurable in common units which allow them to be compared with each other; comparing, say, 'the value of human life lost due to climate change with consumer items or money or other lives'" (Spash 2006: 259).

The closest Gardiner comes to acknowledging this point is his comment that "since the idea is that more consumption can always make up for environmental degradation, the claim surreptitiously implies that suitably bigger economies (and domes) are definitely available to us, and can be brought to bear very quickly". In the light of climate change, Gardiner adds, "such claims are question-begging, and so dangerous" (263). But wherein, more exactly, does this danger reside?

To raise this question is to point to a deficit in Gardiner's discussion of discounting. His criticisms of how Nordhaus in particular conceives of discounting are partly conceptual, partly ethical; and on this basis, he defends Stern against Nordhaus's objections to his 2006 Review. As Gardiner mentions, Nordhaus argues that "Stern's rate is not ethically compelling, given that future people will be richer" (278). But will they?

Whether or not future people will be richer than present ones is neither a conceptual issue nor an ethical question, but an empirical one. As such, it concerns the relationship between the material wealth produced by a growth-oriented economy and the natural world within which its consequences play out. Since the wealth produced by current trajectories involves overconsumption and exploitation of natural resources (to stick to the terminology of mainstream economics), its "unintended" consequence is degradation of the material basis on which economic activity – production as well as consumption – depends, and continues to depend.

The empirical fact of such ongoing degradation does not sit well with the basic premise of mainstream neoclassical economics as well as political theory; in fact, it flatly contradicts it. The premise is that *each new generation will be better off than the one before*. John Rawls, the most important political philosopher in the Western world the last 50 years, following his *A Theory of Justice* (1971), is a key proponent of this premise, captured in the trust he puts on "favourable conditions", that is, conditions conducive to wealth and its fair distribution in the future as well as in the present, the wealth necessary to meet what he stipulates as the basic needs to which everyone is equally entitled, regardless of when they live.

Notwithstanding the disagreement between them, both Nordhaus and Stern postulate an *indefinitely growing economy*, from which it follows – as Dale Jamieson observes – that future generations will be richer than the present generation (2014: 131). It is as if giving up on the assumption that future generations will be – economically, materially – better off than we are is a non-option, a though than cannot be entertained. Why? Would giving it up threaten the very foundations of the dominant theory in both fields, economics and liberal political theory alike?

I believe the answer is yes, and I shall say more about why later. As for where Gardiner stands, he at one point concedes that "future people may not be richer" (285). Why? Gardiner writes that "many of the world's poorer people in 2050 or 2100 may be better off than they are today, but still worse off than current Americans or Europeans" (285). What is more, and much more serious, is that "if climate change has severe effects on matters such as food, water, disease, and the regional economies, then many people in the future may be worse off than people now. (Indeed, some will simply be killed.)". Therefore, there is "no reason to bill these people at a higher tax rate", as Bjorn Lomborg, a supporter of Nordhaus, suggests (285).

This is passage worth paying close attention to. Climate change, we are told, may so undermine the regional economies of many people as to not only make them worse off than now, but even kill them. The cause given for this alarming state of affairs, giving the lie to the idea that each new generation will be better off, is climate change, used for simplicity as a catch-all phrase for humanity's impact on the natural world, an impact that may prove disastrous to the human societies affected by it.

Yet if climate change is what causes the widely held assumption about new generations being better off than present ones to no longer hold, the question is what causes climate change. My answer, which should not be controversial, is that what helps drive climate change is essentially that present-day economies are obsessed with growth, growth in production and consumption (measured as Gross Domestic Product, GDP) taken as the means necessary to attain a situation where future people are better off. While this should not be controversial in terms of the fact of the matter, the causality implied – *growth causes climate change, which causes degradation, which causes people to be less well off in the future* – is remarkably glossed over in Gardiner's discussion: it demands to be at the heart of the discussion, but it is not. If he had been keen to address the causality and the blind spots it sheds light on, Gardiner would seize upon the assumption of continued growth in Stern as well as in Nordhaus. When Stern sets out to estimate the damage caused by climate change in the future, his model implies that in 2200 the world will be 12. 3 time as rich as today. As Dale Jamieson acidly comments, "even if Stern's worst-case scenario comes true and climate change causes a 35 per cent drop in consumption, people in 2200 will be 8 times as rich as we are today" (2014: 142).

Presumably one does not have to a non-economist to see that there is something seriously wrong with this model. The main blind spot I have in mind is the failure to explicitly acknowledge and address that fact that the very means by which people caught up in the present economic system go about becoming better off – materially richer – than they are today, whereby this aspiration is projected onto the future, acted upon by each new generation, in fact undermines this very goal. Producing ever more and consuming ever more of natural resources *thereby* being over-exploited, and so depleted and degraded, is not going to enable people to become richer – be it eight times richer – than we are today, but poorer, their lives taking place in a situation not safer than now but much more volatile and dangerous, with less rather than more to satisfy their most basic needs.

This fact should be an elemental truth to scholars of all disciplines engaged in debates over climate change and how to avert it. I am not suggesting that Gardiner would deny it, put in the terms of causality I employed above. But he fails to reflect on its implications for the positions he discusses. Had he done so, he would encounter limits in them – say, in Stern's case – that, for lack of being subject to critical scrutiny, are tacitly part of his own position as well.

Gardiner illustrates how this happens when he returns to the debate over the correct discount rate. "One might", he says, "reject the view that future benefits and costs must receive lower weights than current benefits and costs because one believes that the current generation has a strong obligation to ensure that future people are better off" (293). Taken to articulate an ethical view, such an obligation to future people seems unproblematic, indeed for many people the only correct one. Yet Gardiner goes on to tell us that this view has met strong opposition from prominent figures such as Lomborg and David Pearce. Writes the latter: "Everything would be transferred to the future…pure equality of treatment for generations would…imply a policy of total current sacrifice" (quoted in Gardiner 2011: 294). Whatever the ethics of this claim, it is, as Gardiner observes, in keeping with "standard positive discount rates" in "protect[ing] the current generation from excessive demands" (Gardiner 2011: 294). When exactly are demands "excessive"? It is hard to tell. But the more important point concerns how Gardiner meets the objection against "a total policy of sacrifice" on the part of the presently living. He says that there would be "something morally problematic in requiring current people to endorse" such a policy. And he goes on to contend that "the problem here is with the discounting, not the ethics of rejecting it" (294).

Given the dire realities of climate change and how it affects nature and humans alike, how seriously should we take an argument to the effect that we risk doing "too much", that we demand more than can reasonably be demanded of ourselves, to deal with the damage done? That this is in fact Lomborg's position is unmistakably signalled in a recent book of his, titled *False Alarm. How Climate Change Panic Costs Us Trillions, Hurts the Poor, and Fails to Fix the Planet.*

One could answer that simply because some prominent scholars in the debate share Lomborg's position, it should get attention – just like the opposite danger, namely that we risk doing "too little", deserves attention. Theoretically, that is, we may err in both directions, and so both are equally worthy of consideration. But this is a false symmetry, a distortion of and distraction from the situation we're in, which is that we as a society, persistently and for much too long, do too little to avert the havoc wrought by ongoing climate change. By accepting, for the sake of argument, that there is a symmetry between the two ways in which we may err, Gardiner is lured by the Lomborg camp of the debate to abstract from the asymmetry of "too little, too late" that characterizes climate policies so far.

What exactly does Gardiner mean when his response to Lomborg is that "the problem here is with the discounting, not the ethics of rejecting it"? Is the problem a technical one, as distinct from a practical one, that is to say, is it how economists conceive of discounting as distinct from the real-life consequences of the position they take on it and the policies they recommend?

I find it difficult to understand Gardiner's position here. Concluding his discussion, he writes that the argument for the appropriateness of cost-benefit analysis to which discounting is wedded is "ostensibly non-ideological": it "asserts that whatever the correct ethical theory, ultimately it must be possible to model that theory in cost-benefit terms" (295). Yet as he is quick to point out, there is a clear preference for utilitarianism in models based on weighing costs and benefits; the affinity between the two is widely recognized, giving the lie to the idea that cost-benefit analysis, and the primacy granted it, is agnostic towards ethical theories. Surprisingly, Gardiner does not make this point, but stresses another one, which is that

> cost benefit analysis "can absorb and neutralize any substantive external criticism, because the materials of such criticism can always be accommodated within the approach. If moral information is left out of a given cost benefit analysis, then we can put things right simply by adjusting the costs, the benefits, or the ways in which they are integrated.
>
> (295)

This is a valid observation, and it leads Gardiner to identify two basic claims implicit in cost-benefit analysis, namely that any moral theory might be modelled in mathematical terms, and that any such modelling may appropriately be labelled "cost-benefit analysis". This in turn "raises deep questions about the limits of mathematical modelling, and about the flexibility of cost-benefit analysis" (297).

These are important critical remarks. But they are internal criticisms, raised from the inside of the models – discounting, cost-benefit analysis – and not from the outside. Perhaps Gardiner believes that to be taken seriously by the world's leading climate economists, he needs to accept their conceptual framework and its tacit epistemological presuppositions. In my view, that is to concede too much ground, with the result that the sort of criticisms aspiring to get a hearing in the scholarly community involved are superficial rather than profound; they lack clout. The effect, then, at best, is that those committed to the models in question seek to modify – improve – them so that the criticisms no longer hold.

1.6 Translating ecological phenomena into economic entities

It is useful to compare Gardiner's way of discussing the limits of the approach shared by the rivals Nordhaus and Stern with that pursued by Clive Spash. Generally speaking, Spash asserts, "the model is more important than the observed nature of the world" (2006: 265). Thanks to "overconfidence in the standard economic models", he continues, economists miss out on the true nature of the world, of empirical reality (264). These are big words, not used by Gardiner, and requiring an explanation.

Spash invokes the pioneering work of Georgescu-Roegen, a founding father of ecological economics, alluded to above, to set out a less internal, yet much more fundamental critique than that found in Gardiner. Georgescu-Roegen seized upon the "mathematisation" of economics, in particular what he termed "aritmomorphic concepts" of the type being routinely, yet inappropriately, "applied to dialectical concepts and qualitative aspects of reality" (2024: 145). His concern, then, is both epistemological and ontological; it is about the fit, or lack of such, between theoretical models and the entities in the real world – be they social, biological or physical – they aspire to describe as correctly, as true to their nature, as possible. What remains important is Georgescu-Roegen's sharp eye for what "quantification can and cannot achieve because it assumes phenomena are discrete and distinct" (2024: 145). In other words, what is problematic is the epistemology and ontology implicit in methodological individualism, the philosophical view that what "really exists" as entities in their own right, irreducibly so, are entities in their being as individual such, say, as so many members of the same category or species; individuals whose interactions with each other, and within clusters or wholes, are to be explained by invoking the basic drives (or needs, or dispositions, or preferences, etc.) of each individual as such. Individuals are what elementarily exists, the way they connect with and relate to each other being regarded as more or less contingent, variable, and context-dependent, whereby the whole – community, species, etc. – they help constitute is ontologically secondary not primary, "artificial" not "primitive", to recall the terms used by Thomas Hobbes (see Wolin 2004).

In later parts of this book, I shall have more to say about why the role played by methodological individualism – not only in standard economics, but in political theory of the kind exemplified by Gardiner – is vital to understanding why climate change has not been sufficiently addressed, in theory as well as in practice. The idea that only discrete individual entities exist, so that whatever extra-individual wholes they help constitute can be explained bottom-up, as it were, is wedded to the postulate, going back to Kepler, that *quantity* is the one essential and truly knowable aspect of reality. Accordingly, the cognition that allows for knowledge in the strict sense consists in measurement and comparison of measurements (see Jonas 1966). In this way, method gains primacy over subject-matter, epistemology over ontology: the formal-mathematical, and abstract and universally applicable requirement of *measurability* comes to define, indeed to exhaust, the domain of the knowable, excluding from scientific knowledge and true existence alike the qualitative aspects of life in its manifoldness.

This philosophical background is relevant to Georgescu-Roegen's assertion that "modern economics has ignored even the most fundamental qualitative aspects of the economic process. Why have standard economists failed, for example, to perceive the essential difference between 'fund' and "flow"?" (quoted in Spash 2024: 145). The fallacy of substitutability, mentioned above, is a case in point, giving rise to various "fantastic conjuring tricks", assuring us that "we must not worry about the exhaustion of natural resources, since with the aid of the ubiquitous Cobb-Douglas production function it can be proved that capital equipment can be substituted without limit for those resources" (quoted in Spash 2024: 50).

Georgescu-Roegen's criticism highlights what Gardiner refers to as *ad hoc* adjustments introduced to protect the underlying model against being proven wrong, rendering its "flexibility" scientifically spurious. Again, this is a valid, Popper-inspired, criticism as far as it goes. But it fails to raise the deeper question about the nature of the *qualitative* aspects of the economic process that are at the core of Georgescu-Roegen's critique. The contrast with the latter shows up Gardiner's critique as a methodologically oriented one, as distinct from an ontologically oriented one.

To appreciate what is at stake here, consider *ecosystem services*. This concept did not exist at the time of Georgescu-Roegen's writing. But in the last 20 years it has become among the most influential buzzwords in climate economics, very much in keeping with its key assumptions at that. The core idea is simple: qualitative processes in the natural world, such as the pollination involving bees, crucial to agriculture, can be translated into a quantitative calculation, enabling us to put a price tag on the "service" in question (e.g., pollination, or water supply) apt to capture its value – read: economic value – for us, as well as to settle the question of how much I, as a consumer, and we as a society, are willing to pay for that service. This in turn enables a whole series of trade-offs, premised as they are on the idea of their complete measurability in quantitative terms, that is to say, in terms of money. In this way, what starts out as a feature of or a process in the natural world, qualitative at that and involving various non-human creatures (or agents), becomes transformed into so many distinct quantitative entities, each with a presumably "correct" price tag, corresponding to its value as a commodity on a market.

The crude way to put this transformation is to say that *ecological phenomena are translated into economic entities*, whereby this very transformation is what allows us to recognize their value – so that prior to being subject to this transformation, ecological processes are "empty" as far as value is concerned, since in a valuational vacuum, virtually as well as metaphorically a "no man's land". The upshot is not only that value is a matter (function) of human attribution rather than residing in non-human, natural processes as such. More fundamentally, it is that value in the form of market value is the only form of value that truly counts, that is apt to make a difference in terms of what is worthy of protection, of going on to exist. There must be a demand for it, reflecting and born of our human need for it, interest in it, preference for it – for want of that component, the human-centred one, and value is simply not forthcoming: for lack of us and our taking-an-interest in it, it is valueless.

There is stark contrast between conceiving value as originating in us humans, and in us alone, hence being something we determine whether we are "willing" to confer to non-human entities, and conceiving value as part of the fabric of the world, encountered in so many non-human species and life forms, hence as originating and flourishing *in re*, in say, the activities of pollinating bees, of growing trees, of thriving rivers and waterfalls, etcetera. The anthropocentrism of the former has been hegemonic in Western culture since the seventeenth century, advanced by such different philosophers as Descartes and Kant, and it is no surprise that it is a key premise in mainstream economic theory to this day, "ecosystem services" being its recent product. The depicted transformation from value as in the world itself, unfolding in nature's various organic processes, to value as solely for us humans and thanks to us reaches its logical endpoint in the statement that "you cannot manage what you cannot measure". Although the contrast made between the two philosophical views on may sound abstract and esoteric, as *acted upon* in the guise of ecosystem services it is pregnant with enormous real-life consequences for the thus "valued" entities in question. Defining an entity as "having value only insofar as it has a productive, service-providing, role to play in the economy" (Spash 2024: 171) is not only a sign of the arrogance of anthropocentrism towards everything other-than-human. More importantly, it gives the green light to policies – practices – where all sorts of natural entities and processes that *fail* to prove their utility for humans in the given economic sense are thereby void of value, and so free for all to exploit with impunity.

Besides being ethically scandalous, the irony of arriving at this endpoint should nor escape us. The attraction of ecosystem services as a concept promising to help tackle climate change is based on the idea that it will be a much-needed *corrective* to policies that have failed to prove sustainable: when we recognize the true value of the goods – services – that nature provides for us, we will be more prepared to do what it takes to *protect* them from degradation. Yet when put into practice, this protection is precisely what is lacking. For something to be *conditional* on human needs and interests as currently perceived, and given a price tag as a result of its importance to producers and consumers within a market economy, secures no protection at all. As long as the notion of natural entities' "principal" substitutability prevails, and so the idea that an X being depleted or going extinct here can be replaced – compensated – by its equivalent somewhere else, enabling their trading and off-setting on a global scale, the practices set in motion may well ruin the natural basis on which the economy remains dependent, like it or not. Instead of securing protection of the value provided us by nature, the opposite may well be the result, namely its destruction as a result of being at the mercy of narrowly human-oriented priorities as stipulated in purely economic terms.

1.7 The role of corruption

I turn now to a feature in Gardiner's book which is truly laudable, namely his inquiry into the importance of corruption. Having established that we pass the buck onto the poor, the future, and nature, he observes that we "face strong temptations to do so" (301). Accepting the reality of climate change, difficult as it is, is not enough, since "our resolve remains vulnerable to corrupt mechanisms of persuasion". Therefore, we need to protect ourselves against rationalization, self-deception, and moral manipulation (301).

Corruption in the sense intended by Gardiner takes many forms, all of them with the same result: inaction. Failing to act when one should have done so, be it for ethical or pragmatic reasons, is a common feature in human agents and in all sorts of circumstances; it is human, all-too-human. Nevertheless, some situations are so serious in terms of the dangers they involve that the failure to act is conspicuous, thus demanding an explanation.

Clearly, climate change is a case in point. Given the available information about the crisis and what causes it, and given the resources available to solve it, why is it that the required action is not taken? The answers typically given are well-known and need not be listed here. Corruption is not among them.

Gardiner notes that "corrupt" has two core meanings: "distorted", as in "altered for the worse", and "immoral or dishonest", as shown by "the exploitation of a position of power or trust for personal gain" (303). Yet he acknowledges that not all corruption is for personal gain, nor does it always involve the exploitation or taking advantage of a superior position. In fact sometimes a group or individual is corrupt even if "has no power and no one trusts it" (304). Nonetheless Gardiner considers the standard definition to suit the case of climate change quite well. Accordingly, in the global storm the main threat is that the rich will take unjust advantage of the poor; in the intergenerational storm, earlier generations will do the same to later generations; and in the ecological storm, the "kick the dog" scenario suggests the further exploitation of non-human nature by humanity (304). The threat posed by climate change insofar as it shows us caught up in a perfect moral storm

> is acute, even potentially fatal, because of the severe nature of some of the asymmetries of power, and because those who are damaged by them – the poor, future generations, animals and the rest of nature – are poorly placed to defend themselves against it.
>
> (304)

But how common is corruption? Is it something we can do without in our everyday dealings with others, becoming activated as it were only in rare cases where much is at stake? Gardiner turns to Kant to suggest that we cannot be so easily reassured: we just can't rely on our normal, common-sense judgments. He quotes Kant's stern yet realistic assessment: "There is something splendid about innocence; but what is bad about it, in turn, is that it cannot protect itself very well and is easily seduced" (quoted in Gardiner 2011: 306). It is part of our makeup as humans that we are inclined, for reasons of selfishness, to engage in rationalizations seeking to undercut or weaken what moral prescriptions demand of us, so as to reconcile them with their own not-so-moral aims. More subtly, this means that we are engaged in *an attack against the foundation of morality*, an attack that "seeks to transform the moral claim itself", thereby perverting our understanding of its status and content (307). The intensity and recurrent nature of this sort of attack stems from the fact that "moral claims are in some sense uncompromising", which is to say that they forbid us to accommodate other sources of motivation (307). It is precisely the way in which moral demands on us crave being *trump*, putting all extra- and non-moral inclinations into second place that is experienced as intolerable.

Over thirty pages, Gardiner develops his understanding of corruption by referring to Jane Austen's novel *Sense and Sensibility*. "Our interest", Gardiner tells us, "lies in how easily John Dashwood [the novel's main protagonist] moves from accepting a serious and apparently unassailable moral commitment to help his stepmother and half-sisters into dismissing that commitment almost entirely." The relevance of Austin's story lies in how it illustrates "just how seductive and familiar the devices of moral corruption are, and how vulnerable we are to them" (310).

There is no need to go into the details of Austin's portrait. It suffices to note the most important similarities that Gardiner sees between the scenario depicted by Austen and that present in the perfect moral storm of climate change. In both cases, he says, the broad outlines of what to do seem clear enough. In line with Kant's emphasis on corruption's essence being an attack on the status – trump – of the moral claim, Gardiner identifies four distinct moves: First, exaggerating the negative consequences of taking the action demanded by the claim, rendering them "excessive burdens". Second, invoking prior entitlement to the effect that the idea that I am obliged to assist others is dismissed. Third, invoking competing special relationships so that those benefited by heeding the moral claim are deemed as less important than others. And fourth, suggesting that no person "in his right mind" would take the moral claim seriously.

In sum, these moves undermine the agent's willingness to act upon the moral claim by throwing into doubt everything that makes it into such a claim in the first place, hence overriding other considerations and priorities. Although Gardiner does not use the distinction, the assault on the moral claim is partly fact-based, partly moral. It is fact-based in dismissing the claim for its lack of realism: "we must recognize the way the world really is", showing up heeding the claim as an act of foolishness (320). And it is moral in picturing the action demanded as unjustified since excessive, amounting to a sacrifice, to giving up something to which the agent is entitled. This being so, refusing to take the moral claim seriously is not proof of egoism; instead, it is a morally sound standpoint against misplaced altruism. If anything is worthy of being condemned, then, it is not the refusal to respect the moral claim but the lack of realism as well as character revealed in giving in to it.

Adapted to the case of climate change, the moves made by the critics of the moral claim in Austen's story has the effect of undermining the notion that we owe something to future generations and non-human nature. Appealing to the right to demand mutual benefit is key here. There must be a kind of balance, if not in the short run then in the long one, between what I give – or give up – and what I get back; most fundamentally, there must be the possibility of *reciprocity*, ensuring that the moral claim not turn into a completely one-sided affair, solely focused on what is demanded of the agent, and demanding nothing in return from those at the receiving end. In other words, "climate action must be a 'win-win' situation for all concerned to be viable" (323).

While the notion that reciprocity is a necessary component of a moral demand is often explicitly stated (more on this below), Gardiner highlights the psychologically more subtle move caught in Austen's appreciation of how the agent is vulnerable to shifting his focus inward, from the concerns of those he is morally obliged to assist to his own concerns, modifying the moral demand so that it comes to be seen as "something to be traded off against his other, more personal priorities, rather than something with a distinct status" (323). The unconditional status of the moral claim, marking it as something that permits of no compromise, is undermined when it is relativized by being drawn into comparison with other concerns, less moral ones at that, but weighing heavily on the agent as more "personal".

Though Gardner doesn't say so, there is a slippery slope involved here from being concerned with the morality of one's other-directed actions to being preoccupied with the question whether one is "authentic", true to one's own values, in heeding the moral claim – *shifting the jury* of one's actions from others (or society at large) to oneself, making my relationship with myself the supreme court of judgment. This is notion with a long history in Western philosophy, going back to Socrates' proposition as formulated in Plato's *Gorgias* that: "It would be better for me that multitudes of men should disagree with me rather than that I, *being one*, should be out of harmony with myself and contradict me" (quoted in Arendt 1978: 181). Of course, Socrates is taken to have thought that this removal from the "outer" to the "inner" perspective would be conducive to moral action, anchoring it in the agent's relationship to himself, the one relationship from which escape is impossible: in killing somebody, I condemn myself to spend the rest of my life with a murderer. Therefore, acting immorally helps create an inner disharmony I cannot possibly, reasonably, want to bring about.

Such shifting of the jury from the outside to the inside, from society to oneself qua "inner tribunal", may seem conducive to moral action in that it gives pride of place to autonomy and independence on the part of the agent, as well as highlighting his or her "authenticity" as conceived by existentialist philosophers from Nietzsche and Kierkegaard to Heidegger and Sartre. But how sure can we be that standing one's ground as against society – be it the "tyranny of the majority" (to recall John Stuart Mill's analysis in *On Liberty*) – is tantamount to siding with what is morally right? Can it not be the other way around – that is, that the agent gives supremacy to coming forward as "independent-minded", and not to be swayed by what others expect, as a trick to escape the claims of morality that Kant talks about? This is clearly Gardiner's message. However, as we shall see in Chapter 4, whistleblowers demonstrate that giving primacy to one's integrity, and thereby risk being declared an enemy by the powers that be, can indeed be what is required for morality to survive in troubled times. There is no one, universally valid answer to this; it depends on the circumstances, on the moral merit – or lack of it – of what "society" expects from the individual agent.

Having looked at the types of attack to which the authority of the moral claim can be subjected, what is it that Gardiner regards as particularly relevant in the case of climate change? The analogy between Austen's portrait of John Dashwood and Gardiner's topic is "generally unsettling", he says. Both involve serious asymmetric vulnerability, where the agents with a moral duty to act "not only suffer little or no negative consequences from a failure to act, but also stand to benefit from that failure"; hence in both cases the central agents are "ripe for moral corruption" (336). Although none of the specific objections raised to dispute the moral claim look very compelling at first glance, "when advanced together they become more difficult to resist" (337). This, then, is essentially why Austen's narrative succeeds in exposing the dynamics in play in a "perfect moral storm" (338). Gardiner concludes his discussion by asking: "In an age dominated by sound bites and intense political partisanship, by limited engagement and short attention spans, by superficial media and spin, are we not also vulnerable to such a barrage?" (338).

Gardiner's answer to his own question is an obvious "yes", so obvious that he doesn't need to state it explicitly. It simply follows from the overwhelming force of the moves aimed at undermining the moral claim's authority to guide our actions.

But why exactly are we so easily influenced by these moves? Why do they prove virtually irresistible? Is it a matter of human psychology, understood as the set of basic inclinations we all share, whatever our differences qua individuals? If this is indeed where Gardiner thinks the root of the problem – our inability to resist attacks against the moral claim – lies, what is the role played by "the age" in which we currently live? How, more precisely, do the features of the latter – say, culture – help shape our present situation as one characterized by widespread inability to resist the forces of moral corruption? Are sound bites, political partisanship, and superficial media the main factors?

As indicated above, in my view these questions can only be fully illuminated if we interrogate the relationship between individual agents – you and me – and the dominant institutions that help shape climate policies in our society. Do these institutions lead the way? Do they provide the conditions necessary for ordinary people to be part of the solution? Or is the truth a much darker one – namely that the institutions – governments and corporations in particular – are *complicit* in the perfect moral storm Gardiner is talking about?

There is a sense in which suggesting the latter is part of the picture of our predicament Gardiner himself paints. Indeed, had the major institutions been up to the task of averting climate crisis, the crisis had not gone from bad to worse over these last 50 years. But to admit their failure is one thing, hardly controversial; quite another is to investigate and lay bare the causes of the failure.

I have repeatedly pointed out that Gardiner's book comes up short with respect to such an explanation. We could put this by saying that the failure of institutions is a major point of departure – a fact – for Gardiner's whole approach, rather than something he seeks to establish for his readers. Yet the sheer familiarity of the failure – it having been with us for so long, having been observed and lamented so routinely – has the unfortunate effect of naturalizing it, being simply the way things are, and have been as long as present generations can remember. We do not seek for an explanation for a situation experienced as completely normal. Instead we adjust to it, letting realism reign by becoming part of its social fabric, and thus helping reproduce it.

If this captures the sense in which Gardiner, along with his readers, regards it as a simple fact that in our society the institutions are complicit in the perfect moral storm he talks about, there is a quite different aspect to consider as well, one much more difficult to get at. What makes the thought that institutions are complicit a dark one in the sense intended, is this: Can it be that they are themselves sites of corruption? Put more starkly still: Can it be that what the institutions in question demand of individual agents is that they become corrupt in the sense of undermining the moral claim instead of heeding it?

I shall not try to answer these questions here, but will come back to them in each of the book's remaining chapters.

1.8 "Bearing witness" in the form of ideal theory

Towards the end of his book, in a section on responsibility, Gardiner asks: "If action is needed, and the rough shape of the burdens clear, who is responsible for making it happen?" (431). He comments that this question may seem almost impossible to answer. Why? Well,

> it seems plausible to claim that our existing institutions were simply not designed for, and did not evolve in response to, global environmental problems that play out over many generations. Hence, it is unclear who has the responsibility and authority to act.
>
> (432)

Why? Is Gardiner suggesting that insofar as institutions fail, the task and the responsibility for shouldering it is shifted from institutions to individuals, to you and me? If that is so, what are the chances that individuals will succeed in a situation where institutions have failed?

Gardiner does not answer these questions, at least not in the form I have given them. Instead he shifts his focus to the failure of "our ethical frameworks" to be up to the task. He cites his philosopher colleague Dale Jamieson's observation that our current values conceive responsibility in a way presupposing that "harms and their causes are individual, that they can be readily identified, and that they are local in time and space" (quoted in Gardiner 2011: 432). Unfortunately, climate change fits none of these criteria, "so that a new value system is needed" (432).

But to state this is to take us back to where we started rather than pointing a way out of the impasse. We are back, that is, to the truism that "humanity currently lacks the appropriate institutions to deal with global environmental change" (433). Seeking for a way forward, Gardiner tells us that "political institutions normally operate under delegated authority from the citizens" – bottom-up, as it were. With regard to climate change, "the delegation has either not happened, or else has failed to be successful". In both cases, there is "a natural answer", namely that

> the responsibility falls back on the citizens again – either to solve the problems themselves, or else, if this is not possible, to create new institutions to do the job. If they fail to do so, then they are subject to moral criticism, for having failed to discharge their original responsibilities. (433)

Note that there is an ambiguity to the "they" in the last sentence: is Gardiner referring to citizens or to institutions? Many would perhaps assume the latter, but it turns out to be the former, as brought out in Gardiner's follow-up question: "If the world's leaders and institutions are failing to deal with climate change, the average person might ask, how does that suddenly become *my* problem?" (433). Admitting that this answer may seem startling, Gardiner reassures us that "it is a traditional one in political theory". Indeed, this model of political responsibility "is built into the foundations of democratic thinking and institutions more generally, as a natural consequence of their basic rationale". He adds that "the whole idea that citizens might be politically responsible for the behaviour of their institutions is in some respects a radical and demanding one" (434). Presumably seeking to comfort those who may find this notion too bold, Gardiner states that success on the part of institutions "breeds the elimination of responsibility at the individual level. The better the rest of the system is at discharging responsibilities on behalf of individuals, the fewer direct demands such responsibilities make on the individual" (434).

I have quoted these passages so extensively because I see them as remarkable. It seems odd to suddenly talk about how "success" on the part of institutions help alleviate responsibility at the individual level, since what

Gardiner sets out to address is the all-too-real *failure* of institutions we are confronted with. More important though is the notion that the relationship between institutions and citizens is a kind of zero-sum game in that the more fully institutions handle responsibilities, the less they demand of the individuals – and vice versa. How well does this sit with the idea that citizens be held "politically responsible for the behaviour of their institutions"? Do current conditions at all facilitate that sort of responsibility on the part of individuals?

To be sure, Gardiner's model is a tall order, theoretically as well as practically. We – ordinary people – are spoken about and taken seriously as *citizens*, with the obligations as well as rights that come with being such in a democratic political order, namely the liberal one that Gardiner, like Rawls before him, takes as his framework. This political order is demanding on both the citizens and the institutions, necessarily so, since this is a hallmark of democracy as distinct from other kinds of regimes. Yet the thrust of Gardiner's argument in describing the "challenges" posed by climate change as a case of "a perfect moral storm" comes from the admission that *both* elements are overwhelmed and therefore unable to rise to the occasion: individuals taken as citizens, as well as the institutions in our liberal political order.

How to respond to this situation of current society being doubly overwhelmed? Would it be a good idea for the political philosopher to modify his model so that it becomes less of a tall order? When does the gap between what the model demands and what its two main parties – citizens and institutions – are capable of doing become so big that the model ceases to be helpful? Is there not a risk that the wider the gap, the more ideological the model?

More fundamentally still: What are the conditions for enacting the role of citizens in present society? Do current institutions and the way they operate foster what it takes to become citizens in the political sense intended, namely mature, responsible adults committed to the common good? Or is it the case that the cultural preconditions for citizenship have been eroded, turning us into consumers from early childhood? In *Consumed: How Markets Corrupt Children, Infantilize Adults, and Swallow Citizens Whole* (2007), Benjamin Barber offers a trenchant diagnosis of "an epoch where shopping seems to have become a more persuasive marker of freedom than voting" (2007: 37). In *One Market Under God: Extreme Capitalism, Market Populism, and the End of Economic Democracy* (2001), Thomas Frank describes

> how big changes are made by market forces beyond our control, but still we must make our own changes to be in compliance with the big ones. We have no say in the mater and yet we are responsible for our own failings;

a situation such that in corporations "talk of empowerment, participation and reengineering is followed automatically by an intensification of management demands" (2001: 244, 246). And in *The Perverse Organisation and its Deadly Sins* (2008), Susan Long, taking her cue from Joel Bakan's work cited earlier, examines the nature of perversion as a basic human phenomenon and how the dynamics of perversion are presently being established in organizations, tracing the movement from a culture of narcissism (Christopher Lasch) towards elements of a perverse culture where individual pleasure is sought at the expense of a more general good, "often to the extent of not recognizing the existence of others or their rights" (2008: 5, 15). The three books mentioned are a random choice; plenty of others would serve to capture the same cultural trend. Note that they are all written *before* Gardiner published his book – he could have paid attention to their analyses and allowed them to weigh in on his own, but he does not. Note also that the trend's wider cultural and political implications – say, for such an issue as confronting climate change – have become much clearer for all to see *after* these books were written: witness the rise of Donald Trump.

Gardiner is on a different page. "Perhaps", he says, "existing institutions and theories must be radically reconceptualized to reflect new global and ecological realities", adding that "it seems wise to acknowledge that climate change involves issues which current political institutions and theories do not seem designed for, nor obviously well-equipped to handle" (435). There is something underwhelming, smacking of anti-climax, to using phrases like "perhaps" and "it seems wise" in describing the mess we're in, the mess the book aspires to do justice to in terms of its severity. What does Gardiner's penchant for understatement betray? That he greatly underestimates the nature of the crisis?

One could answer that Gardiner simply adheres to the conventions of the school of thought he belongs to – political liberalism as shaped by Rawls. This being so, the complacency he identifies as damaging to efforts hitherto made to confront the problem of climate change is part and parcel of his own vantage point as well, only he has difficulties recognizing it there, as opposed to in so many others.

Yet content is more important than what is betrayed by style. Gardiner ends his book by observing that "the project of ideal theory seems pressing". Indeed, "whatever one thinks about the ethics of the transition [required in a situation of a perfect moral storm], it seems clear that ideal theory matters" (435). His adherence to ideal theory is expressed in what I spoke about as a tall order with regard to citizens and institutions alike. He is aware that "some may be pessimistic" about the ability of either to deal with climate "in anything like an ethical way". Yet for what it's worth, he "would resist this pessimism" (437). On what grounds, exactly? Is he busy reassuring his readers, or is he reassuring himself, or both?

Both, as far as I can tell. Resisting pessimism may be an article of faith. It may spring from a human, all-too-human need to believe that things will work out for the better, even if the future looks bleak. Pessimism, on this view, would not be helpful but only undermine our efforts, weaken our resolve, and cause apathy. Rightly understood, Gardiner asserts, our present predicament is precisely the moment for ideal theory. Why? Because it is suited to "the task of *bearing witness* to serious wrongs even when there is little hope of change" (437). By being "ideal", this kind of theory is important in that it helps "remind us that what we do now falls far short of any morally defensible goal" (437).

In other words, we ought to do better, and ideal theory of the sort Gardiner advances is meant to boost our effort to be part of the change that we want to see in the world. But how convincing is the view that the contrast between what we do and what we should do will actually motivate change? David Weisbach asserts that "the ethical arguments that have been made about climate change add little or nothing", proposing solutions that are infeasible. Ethical arguments of the kind put forward by Gardiner, Weisbach continues,

> require high-emitting nations to enter into treaties that make them worse off. Agreeing to a workable treaty will be difficult even if it makes everyone better off. If an agreement will make high-emitting nations worse off, it will be impossible.
>
> (2016: 242f)

Striking a note of realism as opposed to idealism, Weisbach simply does not believe that the solutions to climate change are "centrally about ethics". The problem is caused by the use of fossil fuels, in a world where "virtually everything we do relies on fossil fuels" (2016: 243). If ethical arguments a la Gardiner come up short, where to look for a way forward? Well, to "more efficient solar and wind power, better transmission grids, batteries that hold more energy, new designs for vehicles, treaty design and enforcement mechanism, and so forth" (2016: 244).

There is nothing new under the sun about Weisbach's position, stressing the role of new technologies and market mechanisms, and making the case for pragmatism in a world ruled by the power of self-interest, like it not. In his rejoinder, Gardiner asks us to resist the shunning of ethics. Far from mere hand-waving, he reiterates his conviction that "the key worry is not that ethics may require us to be angels, but that we may (rightly) be remembered as something far worse" (Gardiner 2016: 255).

For my purposes, the debate between Weisbach and Gardiner is instructive for several reasons. For all its "realism", Weisbach's plea for prioritizing technological change over ethical values must be seen for what it is, simply echoing and replicating the strategy that has been promoted by politics as usual ever since climate change became an issue some 40 years ago. Appealing to self-interest may be attractive as a way to reject the idealist folly of ethical theories. Yet this very appeal, ostensibly the strength of hard-headed realism, has dominated how the crisis has been treated all along, and so proved part of the problem not its solution.

Moreover, the claim that an agreement, if reached, will "make high-emitting nations worse off", and therefore be "impossible", is premised on the assumption that the economy is trump – "worse off" is clearly meant in purely economic terms. In reality, however, and to echo Nicholas Georgescu-Roegen and Herman Daly (2007), the economy is not a separate domain from ecology but a subsystem of it, dependent on the integrity and intactness of various ecosystems: no planet, no business. If anything, what constitutes a folly at this point in history is to persist in thinking that the economy can continue to grow in an ecologically degraded world. To the extent that there *is*, in today's world, a conflict between economy and ecology – caused by the former's ongoing degradation *of*, and *within*, the latter – the solution cannot be to continue allowing for the primacy of the destructive party. So, if cutting emissions is considered to make a nation worse off in a situation where doing so is the only ecologically wise, and absolutely necessary, thing to do, this speaks to the heart of the matter: within the constraints and imperatives of present-day capitalist economy, climate change will not be "solved" but exacerbated. This is not a philosophical, let alone ethical argument; it is well-documented empirical fact.

As for Gardiner, his reasons for rejecting Weisbach's trust in the long-reigning strategy of delegating the solution to the interplay of self-interest, new technologies and the market place are sound. But what about his own position?

While it is tempting to agree with Weisbach on one point, at least, namely that Gardiner exaggerates the centrality of ethics in a world run by self-interest, in my perspective a more precise assessment is that Gardiner ends on too defensive a note. Recall his statement: "Although one purpose of ethics is to guide action, in my view it also plays a role in *bearing witness* to serious wrongs even when there is little chance of change" (2016: 255). To be sure, it does seem unrealistic, or downright impossible, that ethics will help bring about the change required. But if this is so, why? Would of different kind of ethical theory than the one favoured by Gardiner have better prospects?

There can be no definitive answer to this question. What I can say, though, is that the absence of a sociological framework that does justice to how individuals and institutions interact once again weakens Gardiner's approach. The individuals – you and me, supposedly – expected to embody the principles and values that Gardiner regards as crucial to an appropriate ethics of climate change are painted in too abstract a manner, detached from the wider cultural context where action takes place. A main task for the chapters that follow is to provide that context.

Bibliography

Arendt, Hannah, 1978. *The Life of the Mind*. London: Secker and Warburg.

Bakan, Joel, 2005. *The Corporation. The Pathological Pursuit of Profit and Power*. London: Constable.

Barber, Benjamin, 2007. *Consumed: How Markets Corrupt Children, Infantilize Adults, and swallow Citizens Whole*. New York: Norton.

Bauman, Zygmunt, 1998. *Globalization. The Human Consequences*. Cambridge: Polity Press.

Dalby, Simon, 2009. *Security and Environmental Change*. Cambridge: Polity Press.

Daly, Herman, 2007. *Ecological Economics and Sustainable Development*. Cheltenham: Edward Elgar.

Frank, Thomas, 2001. *One Market Under God: Extreme Capitalism, Market Populism, and the End of Economic Democracy*. London: Secker and Warburg.

Gardiner, Stephen M., 2011. *A Perfect Moral Storm. The Ethical Tragedy of Climate Change*. Oxford: Oxford University Press.

Gardiner, Stephen M., and David Weisbach, 2016. *Debating Climate Ethics*. Oxford: Oxford University Press.

Georgescu-Roegen, Nicholas, 1979. *The Entropy Law and the Economic Process*. Cambridge, MA: Harvard University Press.

Harvey, David, 2005. *Neoliberalism. A Brief History*. Oxford: Oxford University Press.

Heidegger, Martin, 1962 [1927]. *Being and Time*. Oxford: Blackwell.

Hylland Eriksen, Thomas, 2016. *Overheating. An Anthropology of Accelerated Change*. London: Pluto.

Hylland Eriksen, Thomas, 2024. *Det umistelige. Fra global ensretting til et nytt mangfold*. Oslo: Aschehoug.

Jamieson, Dale, 2014. *Reason in a Dark Time*. Oxford: Oxford University Press.

Jonas, Hans, 1966. *The Phenomenon of Life: Towards a Philosophical Biology*. Evanston: Northwestern University Press.

Jonas, Hans, 1984 [1979]. *The Imperative of Responsibility*. Chicago: University of Chicago Press.

Kant, Immanuel, 1959 [1785]. *Foundations of the Metaphysics of Morals*. Indianapolis: Bobbs-Merrill.

Kemp, Herve, 2008. *How the Rich are Destroying the Earth*. White River Junction: Chelsea Green Publishing.

Klein, Naomi, 1999. *No Logo*. New York: Flamingo.

Klein, Naomi, 2014. *This Changes Everything: Capitalism vs. the Climate*. London: Allen Lane.

Kopenawa, Davi, and Bruce Albert, 2013. *The Falling Sky. Words of a Yanomami Shaman*. Cambridge, MA: Belknap Press.

Lessenich, Stephan, 2016. *Neben uns die Sintflut*. München: Hanser Verlag.

Long, Susan, 2008. *The Perverse Organisation and its Deadly Sins*. London: Karnac.

Mill, John Stuart, 1970 [1859]. *On Liberty*. Harmondsworth: Penguin.

Mirowski, Philip, 2014. *Never Let a Serious Crisis Go to Waste*. New York: Verso.

Oreskes, Naomi, and Erik Conway, 2010. *Merchants of Doubt*. New York: Bloomsbury Press.

Rawls, John, 1971. *A Theory of Justice*. Oxford: Oxford University Press.
Rockstrøm, Johan et al., 2009. "A safe operating space for humanity", *Nature* 461: 472–475.
Spash, Clive L., 2006. *Greenhouse Economics: Value and Ethics*. London: Routledge.
Spash, Clive L., 2024. *Foundations of Social Ecological Economics*. Manchester: Manchester University Press.
Thunberg, Greta, 2019. *No One Is Too Small to Make a Difference*. London: Penguin.
Urry, John, 2014 *Offshoring*. Cambridge: Polity Press.
Vetlesen, Arne Johan, and Rasmus Willig, 2018. *Hva skal vi svare våre barn?* Oslo: Dreyer.
Vetlesen, Arne Johan, 2019. *Cosmologies of the Anthropocene. Panpsychism, Animism, and the Limits of Posthumanism*. London: Routledge.
Wolin, Sheldon, 2004. *Politics and Vision*. Princeton: Princeton University Press.
Wolin, Sheldon, 2018. *Democracy Incorporated. Managed Democracy and the Specter of Inverted Totalitarianism*. Princeton: Princeton University Press.

2 Killing the conversation

The social organization of denial

2.1 Denial: the default position?

In his book *What We Think About When We Try Not to Think About Global Warming: Toward a New Psychology of Climate Action* (2015), the Norwegian economist and psychologist Per Espen Stoknes seeks to "look deep into the psychology of the human response to climate change and show how to bypass the psychological barriers to action" (xvii). He tells us that the book is a product of his long collaboration with his older colleague Jørgen Randers, who rose to fame at the age of 27 when he co-authored *Limits to Growth* in 1972, sparking a heated debate over the thesis that global consumption would overshoot the planet's resources. With his co-authors Donella and Dennis Meadows, Randers tied overshoot to a combination of rapid change, limits to that change, and "errors or delays in perceiving the limits and controlling the change" (2005: 5). That there are limits to growth, since all resources are limited, is a fact to which the world has been blind. This blindness, Randers has claimed since 1972, has helped lead to the situation of overshoot that has only become more serious since his first warning to the public.

Why are we blind to the fact there are limits to growth, and that overstepping them will prove catastrophic? Individuals, Randers tells us, continue to support growth-oriented policies because they believe growth will provide ever-increasing welfare. Government and corporate leaders try to meet this expectation by "doing all they can to produce more and more growth" (2005: 6). Growth continues to be seen as a synonym for, as well as a condition for, things that are good and worthy of celebration, such as development, progress, advance, improvement, and prosperity. Growth being positive and inherently desirable, as it were, its alternatives in the form of growth no longer continuing appear negative. This being so, any message to the effect that growth needs to stop will fall on deaf ears.

Stoknes takes his mentor's key questions as the point of departure for his book: Is humanity up to the task? Are we humans "inescapably locked into

DOI: 10.4324/9781003660743-3

short-termism" (2015: xvii)? While sharing Randers's view about what the problem is and what causes it, Stoknes's stated aim is to prove his colleague wrong: in rejecting Randers's pessimistic conclusion that "the socioeconomic juggernaut [will] run its inevitable course of overshoot and decline", Stoknes wants to come across as an optimist whose message is that the climate paradox is resolvable, since "there is more to humans and cultures than short-termism" (xviii).

What is the climate paradox? While the scientific data regarding climate change are more and more alarming, people in many countries – the United States and Norway among them – seem to care less and less. The known facts are all about reasons for concern, for taking action, urgently at that – and yet the response is the opposite, to choose to look away, "despite the dire facts, or perhaps exactly because of them", writes Stoknes (xviii). Choosing to look away in the sense intended is a matter of being in *denial*; it is "both to know and simultaneously not-to-know" (16). The knowledge is such as to motivate action, yet the response is to refuse to act. Why is that? It is because the reality conveyed by the knowledge is an uncomfortable one, encouraging escaping it in different ways.

Escape takes the form of denial, itself a complex phenomenon. Drawing on sociologist Stanley Cohen's classic study *States of Denial* (2001), Stoknes highlights the difference between active and passive denial. In *active* denial, there is "knowledge of the facts to a certain degree, but they are energetically refuted and refused and rewritten since they don't fit with your values – personal, political, or otherwise" (17). By way of example, Stoknes cites the American officer who commented on the My Lai massacre in Vietnam: "There was no massacre and the bastards got what they deserved." This captures how what is refuted is indirectly accepted in the same statement. In contrast, *passive* denial is "more on the indifferent, unresponsive side"; you "may know climate change exists, but prefer not to care much about it" (17).

Crucially, the sort of denial we talk about in connection with climate change is a strategy chosen one for the purpose of protection. Taking in the facts would be too difficult, too painful. But to what extent is this because of the nature of the facts as such, and to what extent does their being "too painful" say something about me? The answer is that there is no either/or, but a both/and. Why? Because in the case of climate change what is involved is the way in which I as a person and climate change as a fact in the world are *connected*: there is a relationship between the two such that what the knowledge is *about* is a phenomenon that is not indifferent to me. It might be tempting to invoke the saying that "while you may not be interested in war, war is interested in you" as a parallel. But as a person living in today's society, I *am* interested in climate change, ineluctably at that, since it shapes the world I live in, shapes it ever more rapidly, changing it dramatically in the course of my lifetime. And unless I plan on escaping to a different planet, there is no way to escape its all-too-real consequences. This very no-exit, non-exemption, is what fuels denial in our time, which is to say: what gives it its truly novel guise.

This is my way of putting it, not Stoknes's. Where do we differ?

Stoknes makes the observation that "there is always a component to denial that goes deeper than likes or dislikes" (17). In other words, we engage in denial when there is something at stake for us, something we experience as important. Stoknes puts it like this:

> The term *denial* is appropriate when the full acknowledgement of what is denied would imply *having* to act upon it. The lifting of denial would result in an emotional shift; it would require both speaking and acting differently. And sometimes it would result in a substantial change of lifestyle, ethics, and identity.
>
> (18)

This captures the sense in which accepting the reality would be compelling: it would dictate action, that is to say, force me to change insofar as this is a reality that cannot go on. In other words, what is at issue is being ready to accept that something that is part of present reality is *unacceptable*: it cannot stand. Accordingly, the lifting of denial comes with a demand: coming, finally, to see reality for what it is, I must act so as to do justice to what I now acknowledge; confronting reality, I need to change, substantially so, as Stoknes points out.

The shift involved, then, is one of perception rather than a shift in reality as such: a shift from regarding present reality as acceptable, hence not calling for change, to regarding it as so unacceptable as to dictate change. This prompts the question: where does this shift in perception come from?

Stoknes remarks that "to what extent this individual shift out of denial is needed for cultural shifts to happen…is shrouded in uncertainty" (18). He turns to some historical examples: the many whites in South Africa in the 1980s who were in denial about apartheid; the cultural denial that preceded the abolition of slavery during the nineteenth century; the US attitude towards the Vietnam War in 1970; and the battle to get people to acknowledge the adverse effects of tobacco smoking in the 1990s.

The examples given by Stoknes have been invoked by many authors before him, especially the abolition of slavery. How illuminating are they with regard to climate change, or to the unfolding environmental crisis more broadly? Stoknes doesn't say, he is content to list them, taking their relevance for granted. To anticipate later discussion, it seems to me that his examples fail to match the depth of how people in today's society are implicated in climate change and the structures that sustain and accelerate it – implicated in a way that was not the case with slavery or apartheid, let alone smoking. What does implicated mean here?

For a start, it means to be part of the problem, partaking in what causes it. Driving a car, eating red meat, going on holiday by plane, buying food imported from different continents, having kids – these and countless other activities make up the lifestyle of an average Norwegian like myself, and each one of them contributes to climate change understood as the accumulated human-caused impact on the environment; an impact that nowadays, in the Anthropocene, has become so comprehensive and profound as to change that environment, the natural world where the consequences of everything we do play out. Since I am implicated by the way I live, and in particular by the patterns of consumption that my everyday activities help reproduce, this very fact raises questions about responsibility and guilt. Am I really to blame for this situation? Aren't the causes of the problem to be located in the overarching structures of society – say, the economic system – rather than in the individual's behaviour? And even if I think that I am indeed part of the problem in the sense that, morally speaking, I should endeavour to change my lifestyle, will my effort be up to the task? If the structures don't change, what can my decision to change accomplish? How to avoid that what starts out as taking personal responsibility ends up being overwhelmed by feelings of futility and powerlessness? Is climate change a personal and moral issue at all? Isn't it rather a political one, residing with those with the power and tools to truly make a difference?

The way I read him, Stoknes's reply is that the question is mistakenly posed as a either/or. The only way forward is both-and, meaning a combination of individual effort and system change. So far, no party has been willing to make the first move: politicians are waiting for their voters to get ready for change, to demand change, while ordinary people are waiting for the politicians to show the way, to assume responsibility for the changes needed on a structural level. The mutual waiting, meaning foot-dragging, has been going on ever since the problem was announced in the 1980s, resulting in an impasse that seems to implicate all parties, regardless of their power or lack of it. The sheer fact that so little is happening, that figures of authority say again and again that the problem is getting increasingly serious, yet having little to show in terms of measures taken, is psychologically demotivating, encouraging the moral corruption we saw Gardiner speak about, turning climate change into a wholly "negative" phenomenon there is a plethora of good reasons to keep at a distance, if not downright avoid – or deny.

"Political scientists, economists, strategists, sociologists, and anthropologists" writes Stoknes, "each have their favourite levels to explain why we disagree, and why so little is happening. None of them really address the role of denial." That is why he will look "primarily into the more basic level" (22). This is not to say that he will not pay attention to "how denialism plays itself out in the social and cultural networks". However, experience has taught us that the overarching levels – international, national, corporate, and cultural – all seem in a deadlock, so that "no major decisions will be made – at least in democratic countries – until vocal and numerous citizens demand it". This being so, "rather than just waiting for the top-down approaches, maybe we could start by understanding apathy and denial at the individual and small-group level" (23).

Whether or not one agrees with this as an analysis of power, Stoknes's approach makes sense methodologically: it is important to explore how denial is being cultivated and communicated among individuals, denial being a feature of individuals, not structures. In other words, to fight denial and its social impact is to confront it in the individuals who embody it in their mentality and behaviour. And as mentioned above, confronting denial with regard to climate change is ethically laden, non-neutral, in insisting that there is an ethical obligation to respond. Indeed, says Stoknes, those who live out their denial "actually support the ongoing human violence to the land and clean air"; to support denialism is to "lend our cognitive capacity, creativity even, to support a destructive cause" (25).

These are harsh words coming from a psychologist who is also an economist; harsh in their moral judgment. I think it's a good thing to be this blunt about the morally dubious nature of engaging in denial, especially in a situation where embracing it has lost the innocence it may have had at a time when the destruction was both less visible and less scientifically proved, say, 20 or 30 years ago. The fact that the destruction Stoknes talks about is ever-growing, virtually everywhere, regardless of where you live, makes denial more contentious, and more demanding of an effort, an uphill struggle as it were.

And yet – denial is in many ways *the default position*, even today, or more to the point: especially today. How can that be?

It has to do with a fact that Stoknes fails to address in its full implications: that the society I belong to, which I am now told is involved in destruction on a potentially catastrophic scale, is the society that is the greatest success story there ever was, in particular if you have the rare luck of being born in Norway, one of the most affluent, tolerant, and harmonious societies in human history. If Stoknes's definition of the paradox his book is premised on is that "if the facts are so compelling, why do they not really register?", the deeper paradox predating it and helping shape it is this one: If the society I am born into is the most successful one in human history, how can it be that it is suddenly to be seen as a threat to the future; as no longer tantamount to prosperity and progress, but to its opposite, destruction? Is being a member of this society a reason for pride and contentment, or a reason for bad conscience and for fearing the future? If so much that is good has brought so much that is bad, to the point of endangering the future, how I am to feel about this society – as a source of shame and guilt? If so, where does this leave me qua individual? Can I remain loyal to this society if I accept, as real, as *undeniable*, the destruction it is causing?

2.2 Socially organized denial

Why do I say that Stoknes fails to appreciate the full implications of the latter paradox? To see why, we must look at the important role played by *cognitive dissonance* in his book. In keeping with the established definition, coined by American psychologist Leon Festinger, Stoknes takes cognitive dissonance to refer to "a situation in which there are conflicting feelings, thoughts, and behaviours", creating an uncomfortable inner tension (61). A paradigm example is smoking: "I smoke. I also know smoking leads to cancer." The fact that these two thoughts – one referring to what I do, the other to what I know – don't go well together leads to tension; and given that this tension is uncomfortable, I seek for strategies to help resolve or remove it. Prominent among such strategies is downplaying, or casting into doubt, one of the two factors making for conflict. I can doubt the validity of the knowledge bit: perhaps I rely upon the wrong authorities, or perhaps the science isn't unequivocal and fully proved; I welcome dissenting voices. Or I can doubt the significance of the behaviour bit: I don't smoke that much; my father has smoked more than I do, and he is still in excellent health at 81; besides, smoking really means a lot to me; and, to invoke the issue of moral authority for a second, who has the right to tell me to stop smoking, being an autonomous adult who takes responsibility for my decisions?

Turning from examples like smoking to climate change, Stoknes mentions the well-known strategies for reducing the tension caused by dissonance: "It's not that serious. Maybe the science isn't settled yet. Maybe the models are wrong" (65). If nothing else works, I can dismiss the whole thing, citing lack of evidence, wild exaggerations of the so-called problem, people committed to doomsday scenarios, etcetera. The reward is obvious: "Weakening the climate message makes us feel better about ourselves and the high-carbon lifestyles we all have" (64).

I take it that we all readily recognize ourselves, and people around us, in these examples. Still, there is something trivial and superficial about them, since what Stoknes sets out to explore is denial in the face of *climate change*. Can the threat posed by accepting the consequences of smoking be compared to the threat posed by accepting climate change? How can I hope to neutralize the fears connected to climate change compared to the fears connected to smoking? To admit that the case for the dangers of smoking is compelling may lead me to solve the tension and get rid of my fear by quitting smoking, be it today. To admit that the case for the dangers of climate change is compelling…would lead me to do what?

This is where the analogies and comparisons suggested by Stoknes fail to convince. They stay on the surface, evading the depth of the issue, that is to say, of how deeply entrenched we are in *a culture* that is pervaded by a mentality and by practices that contribute, causally, to climate change – not in a static sense, but in the dynamic one of amplifying and exacerbating the crisis, and doing so in a situation where the "official" stance on climate change – from the science community, from CEOs, from politicians and ministers – is that it is real, a problem that urgently needs to be solved.

Of course, living in a society where scientific knowledge has one message – the future is dangerous – and the everyday practices another – I am entitled to my present way of living – makes for a conflict and thus tension in Stoknes's sense. Similarly to what we saw in the case of Gardiner above, Stoknes's way of theorizing this tension, and the strategies sought, is to *internalize* them, pointing to an "inner tension" in the distinct individual, as my struggle with – or against – myself, a struggle that aims to remove the tension by casting so much doubt upon one of the two factors – typically, knowledge vs. behaviour – that what I think (or feel) and what I do not longer stand in contradiction. The need for denial stems from this contradiction within me; its aim is to restore harmony where there was conflict in my relationship with myself (to allude to Socrates).

In suggesting that a level of depth is lacking in Stoknes's approach, what do I have in mind?

Stanley Cohen's classic study *States of Denial* is helpful here. Declaring it to be the model for his definition of denial, Stoknes focuses on Cohen's distinction between two major types of denial, active and passive. More important, though, is Cohen's identification of the following three varieties of denial: literal, interpretive, and implicatory. *Literal* denial is "the assertion that something did not happen or is not true" (Cohen 2001: 7). Though not mentioned by Cohen, we may consider climate change deniers – or sceptics – as a case in point. In *interpretive* denial, the facts themselves are not denied but are instead given a different interpretation, taken to mean something else; euphemisms such as "collateral damage" are used to alter, or downright dispute, the meaning of the facts and events. The third category, *implicatory* denial, is the most relevant. What is minimized here is not information, but "the psychological, political or moral implications that conventionally follow" (2001: 8).

Unlike Stoknes, the American sociologist Karie Marie Norgaard pays much attention to implicatory denial in her book *Living in Denial: Climate Change, Emotions and Everyday Life* (2011), based on her fieldwork in a rural town she calls "Bygdaby" (later identified as Voss, close to Bergen, in Norway). As I pointed out in the Introduction, Norway is a good place to study denial because the contradiction between knowledge and action is particularly visible in a country combining, on the one hand, high levels of education, idealism, and appreciation of nature (tied to popular outdoor activities such as hiking and skiing), and on the other, a vibrant petroleum-based economy, employing more than 200 000 people, including subcontractors, To put the importance of the oil and gas industry into perspective,

Norgaard notes that as of 2009 (i.e., the time of her fieldwork), Norway was the world's fifth largest oil exporter and the second largest exporter of natural gas. She observes that in the ten years prior to her stay in Bygdaby, Norwegian oil and gas production increased threefold. In the same period the Norwegian government dropped the goal of capping national carbon dioxide emissions, instead opting to expand petroleum development, and pursuing a strategy for trading emissions rather than reducing them outright. Let me add that after Norgaard published her study in 2011, the government has stepped up production even more, citing the greatly increased demand for gas in particular in Europe following Putin's full-scale attack on Ukraine in February 2022. In a move fitting to this trend, the partly state-owned major Norwegian oil and gas company Equinor has abandoned its much-publicized aim of dedicating at least 5% of its investments to renewable, non-fossil energy (wind and solar); as of 2024, instead of growing, the portion has been reduced to less than 1%.

Voss (or "Bygdaby") is a particularly suitable place to study denial for several reasons. To begin with, it is one of the towns in Norway most well-known for the inhabitants' commitment to tradition, especially to outdoors activities such as skiing and hiking in the mountains. To master skiing is a crucial part of the socialization of children, and the older generations are proud that Voss is a place offering marvellous experiences in, and of, nature, with majestic mountains, vibrant rivers and excellent facilities for alpine as sell as cross-country skiing. The landscape is beautiful, and each of the year's four seasons has its distinct characteristics, from the still whiteness of the snow covering everything in sight during the long winter with its mere four or five hours of daylight, to the strong colours – from light yellow to dark red – of the tree-covered valleys that surround the town on all sides.

When Norgaard came to Voss to begin her fieldwork in 2009, the impact of climate change was becoming more visible each year. The four seasons had lost their stable rhythm, the predictable way in which they alternated, with winters stretching from the first snowfall in November to the end of April, giving way to spring, vibrant with birdsong and creeks thriving for all to see once the snow had melted; four seasons with such variance in temperatures and the sorts of activities allowed for – whereby being outdoors is the one constant, whatever the weather – that they combine to make Voss into a fantastic place to live, so close to nature in all its splendour. More than anything else, the qualities of the natural surroundings made for a high quality of life, inviting people from all generations to participate in and gather around skiing in particular.

Coming to Voss at a time when climate change had visibly begun to disturb the stability of the four-season cycle, causing what we have come to call "disruption", meant that Norgaard expected people's responses to these changes to be an obvious, even paramount, topic in everyday conversation.

It soon turned out she was wrong: far from prompting an open exchange about the changes occurring in their surroundings, such as the winters becoming ever shorter and more unpredictable, with rain in January and other erratic, out-of-joint developments, bringing up climate change "often killed the conversation": "People gave an initial reaction of concern, and then we hit a dead zone where there was suddenly not much to be said, 'nothing to talk about', resorting to "stock" replies, easy answers, or even easy reflections" (2011: 55). Triggering feelings of helplessness and despair, the topic of global warming and its effects, no longer abstract and invisible, but concrete and visible, was subject to evasion, whereby Norgaard herself felt the pull exerted by her conversation partners to move the topic to other concerns, presumably less uncomfortable and contentious ones, to return to a shared comfort zone. She relates how all the educators she met, such as teachers who taught the science in the schools, stressed the need to be always *positive* when discussing these issues, showing a trust in technological innovation and entrepreneurship, reassuring the pupils that the problem is being handled by those in charge.

Norgaard's fieldwork makes for an instructive comparison with Stoknes's approach. "In sharp contrast to psychological approaches to denial", writes Norgaard, she highlights the notion of *socially organized denial*, emphasizing that people's way of reacting to information about climate change "occurs in response to social circumstances and is carried out through a process of social interaction" (2011: 9). While Stoknes tends to focus on the intra-individual, intra-psychic dimension – the "inner tension" discussed above – Norgaard directs her attention at "how features of culture and everyday life make it harder for people to think about climate change and easier for them to ignore it" (2011: 13). The conversations she initiated did not, in most cases, disclose "a rejection of information per se, but the failure to integrate this knowledge into everyday life or to transform it into social action" (2011: 11). At this point, Norgaard refers to Stanley Cohen's notion of *implicatory denial*, describing it as "techniques for avoiding moral and psychological demands" (2001: 9). To avoid these demands amounts to denying the *significance* – the meaning, the consequences – of the reality at hand. Note that in implicatory denial the facts of the matter and "even their conventional interpretations" are conceded, are not as such denied; only "their expected implications – emotional or moral – are not recognized" (2001: 22).

Culture plays a prominent part in Norgaard's approach, comparable to that played by ideology in Cohen's. She employs Ann Swidler's concept of culture as a "tool kit" of shared symbolic resources, a kit that members of Bygdaby would draw upon in the forms of "tools of order" and "tools of innocence" to "create distance from responsibility, to assert rightness or goodness of actions, to maintain order and security, and to construct a sense of innocence in the face of disturbing emotions associated with climate change" (2011: 11f.). Echoing Clive Hamilton's observation, cited in the Introduction, that "denial is due to a surplus of culture rather than a deficit of information", Norgaard's aim is to show how public non-response to climate change and its visible impact in one's home area is "*produced* through cultural practices of everyday life" (2011: 207). When people "actively normalize climate change", they do so by way of taking part in and drawing upon the cultural practices prevalent in their community, practices that provide their lives with shared meanings and common values, a sense of purpose and identity, being a person who belongs and has a role to play in various social activities. Crucially, this means that denial, although observable – manifest – in the distinct individual's behaviour, is a thoroughly cultural and inter-subjective phenomenon, something deeply – and most of the time tacitly – *shared* by the individuals constituting the community rather than something setting them apart and bringing them into overt conflict with each other (to allude to the popular notion of "polarization"). If anything, then, the stance that would allow one individual to stand out from the community would be that of *refusing to take part in denial*, hence of non-compliance, something that comes with a considerable social cost.

The difference between the two approaches can now be set out. Stoknes concentrates on the inner tension in the individual who is caught up in cognitive dissonance, resulting in an uncomfortable inner conflict that needs to be resolved. The main source of the tension/conflict is the contradiction entailed in knowing and not-knowing at the same time; it is about how what I know can be brought into harmony with what I do, and this typically engages me in denial. In Norgaard attention is also given to cognitive dissonance as created by discomforting knowledge about the reality of climate change. But in her perspective, the social relationships and cultural patterns in which the individual is embedded encourage it to evade the contradiction involved in "knowing and not-knowing" at the same time. The impact exerted by culture is such as to encourage social conformity and agreement, not only in conversation – verbally – but in the behaviour constantly interpreted and evaluated by the individual's significant others, that is, family members, friends, neighbours, and colleagues. What Norgaard helps us realize, then, is that culture in the sense defined sees to it that the tension-producing contradiction Stoknes concentrates on – the knowledge/behaviour paradox – in the individual agent, is pre-emptively handled, as it were, so that the shared culture is instrumental in the bringing about of denial. Rather than a property or effort on the part of the individual, *denial is the socially expected behaviour*, the norm, turning the individual who refuses to embrace it into the exception from the rule, an outsider.

The insight Norgaard helps us gain, then, is that culture is *complicit* in denial. Instead of supplying the individual with the symbolic resources needed to *resist* denial, to demand the authorities to tell the truth about climate change and to take tough measures to tackle it, culture in the form Norgaard experienced it in Bygdaby met the deeply human need for *reassurance* halfway, offering a sense of confidence such that the individual would not be "unnecessary" alarmed. The individual's need for reassurance in a situation where discomfort looms – it can't be denied that the climate where I live *is* rapidly changing – is so readily met by the culture shared in everyday life as to make for what is seemingly a win-win situation, rendering it a daunting task for any individual to challenge its hold on people.

For all their importance, these are very general observations about how individuals and the culture they are part of each play a part in advancing denial. Since Norgaard is doing her fieldwork in a Norwegian town, the particularity of the culture at *that* time and place needs to be caught as accurately as possible. It is my impression that Norgaard, living in the United States and staying in Norway for the limited time of her fieldwork, has not quite grasped the extent to which the country's recent history as a major producer of oil has shaped the mentality of contemporary Norwegians.

Norgaard rightly stresses the importance of locating emotions in a political-economic context. For ordinary Norwegians, and not only inhabitants of Bygdaby, there is a strong connection between the country's 50 years history as a world leading producer and exporter of oil and gas, on the one hand, and the economic prosperity and high standard of living this has brought to the population as a whole during that period, on the other. What this means culturally is that utilizing fossil fuels accessed within Norwegian territory is regarded as having been immensely beneficial to "us all". This view is shared across political and geographical divides, and by all generations, although (as we will see later) some generation-based tension has emerged in recent years. Generally speaking, then, what is referred to as the "oil adventure" ("oljeeventyret" in Norwegian, suggesting a fairy-tale come true) is considered a strike of luck for which all presently living Norwegians have to feel grateful, not least to the professions – including deep-sea divers – involved in the early phase in the North Sea, fraught with danger, risk-taking, and bold engineering entrepreneurship, exhibiting a laudable willingness to put the well-being of the country before everything else. While we were simply lucky to find oil in our waters, the story goes, it was the hard work of one generation in particular that helped turn the fossil energy into a financial cornerstone of the modern Norwegian welfare state. Norwegians from all walks of life and all parts of the country are placed in a position of debt to those who devoted themselves so unselfishly to helping advance the prosperity of the whole nation, a sort of debt whose basis is economic but whose implications are psychological and moral, and as such deeply embedded in Norwegian culture.

To appreciate the full force of this particular culture, inseparable from the story about the benefits of oil, we need to note the importance of the notion of *entitlement*, not mentioned as such by Stoknes and Norgaard. Entitlement refers to a sense of what I have a right to do and to be in a broad sense, that is to say, the kind of life I deserve to live, with all the good things that is considered required, both materially (consumption, possessions) and in terms of experiences (traveling to new parts of the world). In exploring denial in present-day Norwegian society, we need to acknowledge, on the one hand, "the general egoism and hyper-entitlement of Western culture in the industrial era", that psychoanalyst Donna Orange talks about (2017: 65), and on the other hand, the role that being a producer of fossil fuels is playing in Norway. We need to recognize how a national culture of indebtedness and gratitude to those (often family members) employed in the oil and gas industry goes hand in hand with the more general (Western) culture of entitlement in Orange's sense, a culture of which Norway is a part, and indeed also was so prior to the onset in the early 1970s of the "oil adventure". Indeed, it is not only that the latter fits with the former; it also helps sustain and amplify it over generations. Given this cultural context, young people who voice criticism of Norway's continued reliance on oil are told that they are ungrateful and disloyal – spoiled children who take their welfare state for granted without paying due respect to its history (I return to this topic in Chapter 4).

2.3 Industrial ambivalence

The constellation of entitlement and gratitude helps illuminate a psychological phenomenon that is crucial in denial concerning climate change, namely ambivalence. Suggesting the term "industrial ambivalence", the British psychologist Renee Lertzman, in her book *Environmental Melancholia*, writes that

> while natural [environmental] damage may be experienced as loss, the causes of the loss (e.g., industry) are also imbued with benefit, achievement or surplus. We partake of the fruits of industry whilst we may experience sadness or longing for the losses it incurs.
>
> (Lertzman 2015: 107; see Connor 2016: 70ff.)

In my opinion a more profound notion than the much discussed one of cognitive dissonance, given pride of place by Stoknes, *ambivalence* refers to simultaneously holding competing affective investments, be it love and hate or gratitude and anger. As Lertzman found in her interviews with residents of Green Bay, an industrialized urban area in Wisconsin, the United States:

> Ambivalence characterizes the mood with which participants tended to relate with the local industry and the local environments. On the one hand, many openly expressed disgust and shock towards the level of local

> water and air pollution, yet on the other hand they quickly shifted modes and contradicted themselves to provide excuses or rationales for the very issues they just reported happiness over.
>
> (Lertzman 2015: 107)

Typically, the respondents would say that industry is an evil, albeit a necessary one, hastening to add that "Green Bay is a good place to live". Environmental health problems that are initially acknowledged are disavowed a moment later, then acknowledged again, and so forth. As a rule, the admission of a blatantly negative feeling (anxiety, disgust, anger, sadness, and hopelessness) towards man-made environmental harms will never be allowed to stand – especially so when the harms occur close to home and are, as such, undeniable since experienced concretely and first-hand.

More psychoanalytically oriented than both Stoknes and Norgaard, Lertzman suspects that the negative feelings and the anxieties they produce are in a sense more true, more authentic, than the disclaimers. What strikes the observer as contradictory and confusing, then, in noting the going back and forth between opposite feelings, is a persistent split in the person's psyche: when ambivalence runs as deep as this, never to be resolved, the price paid is inner conflict. The cohabitation of opposite feelings is no less torturous and exhausting for having an asymmetric rather than symmetric structure, insofar as the "positive" feelings are so much more welcome, culturally and socially speaking, than the negative ones, especially when they express anger or border on despair and depression. But it is not only that such a person will tend to adjust to the expected norms of his group by hiding his "negatives" and making sure to display his "positives" ("I'm sure it will work out"; "This is a wonderful place to live"). He will also engage in self-censorship, spending psychic energy to suppress the negative feelings and thoughts – a part of his emotional life that he not only wants to prevent others from seeing, but does not want to acknowledge to himself either. Feelings that have nowhere to go, be it inward or outward, must be disowned.

There is no doubt that Lertzman's material from Wisconsin, the United States, supplements Norgaard's from Voss, Norway. In both countries and cases, experiencing industry (be it petrochemical, or oil and gas) as both benevolent and brutalizing contributes to a peculiar type of what Lertzman calls "industrial ambivalence" in which "we can be deeply attached to the very practices that are recognized as harmful and even damaging" (2015: 121). Indeed, how could we as individuals living in a place where environmental degradation takes place, accumulatively over the years and generations, *not* respond by being torn between our culture's narrative that "development" is a god thing, allowing us to "have it all" in keeping with a sense of entitlement, and feelings of anxiety and despair when we – simultaneously – cannot but register the often eminently concrete loss that comes with the "progress" and that seems to be the price to be paid, inevitably at that. Lertzman captures the two-in-one like this:

> The Fox River is an ecosystem and a dumping site; it is abject and glittering, a gem or place of trauma. It contains shopping trolleys, PCBs, algae, fish, and mayflies. The river *is* both an ecosystem and a waste site. It is both industrial and natural.
>
> (2015: 121)

So when Norgaard's respondents insist that Voss is a good place to live (despite the seasons becoming erratic, with rain instead of snowfall in midwinter, disrupting all sorts of activities held dear and associated with the place – home – and its cherished qualities), or when Lertzman's respondents say the same about living in Green Bay (despite the river being polluted, fishing becoming forbidden, and swimming no longer attractive), what they say is true. But only partially so. It being only partially true is psychologically non-neutral: it is something that hurts, because something that should not be, evidence of a world gone wrong. Hence the loss involved is a difficult one twice over: painful in itself, subjectively so for objective reasons, yet at the same time socially discouraged, if not downright taboo: it is not a loss that can be socially processed and recognized. It is something *unshareable*, echoing Norgaard's experience that the topic of climate change would "kill the conversation". For lack of public recognition, owing to the absence of a cultural-symbolic space to turn to, the mourning (in Freud's sense) that would permit the individual to work through it by doing justice to the reality of the loss and what caused it cannot take place. Perforce, then, the loss becomes melancholic, being attributed to unclear and to a large extent unacknowledged origins, evolving into what Lertzman describes as "a mood of sadness or vague disappointment" (2015: 127). Whereas mourning, as Freud maintained in his classic *Mourning and Melancholia* (1985 [1917]), makes it possible to move on, even to grow in the wake of the loss suffered, melancholia is more of a dead end in that it sticks, disallowing the person to overcome it and to regain the energy lost.

However, in drawing on Freud, Lertzman should have paid more attention to the difference between the cases of, and causes for, loss that so preoccupied him, being inter-subjective in that the "object" lost was invariably taken to be a significant human other (say, a partner), and the cases under discussion in her own material, where *the natural world*, not fellow human persons, is what prompts my emotional response. The sort of loss that takes place *in nature* is different from the one that affects, or even ends, human relationships, the former pointing to the human–nature relationship in a way that Freud could not anticipate, hence did not theorize, because it was not disrupted – precarious, endangered – in Freud's time the way it is today, in the Anthropocene. Crudely put, whereas nature in Freud's time was considered a trustworthy and stable reality to turn to when things human and social got difficult to handle, always there for me to comfort me, nature in our era has precisely lost – been robbed of – that ontological sort of stability and trustworthiness. Nature itself has become precarious and volatile in an historically unprecedented sense, hence also our relationship to it; to mourn that nature has been altered – harmed – in this sense, because of us humans and our actions and lifestyle, is what "eco-grief" is all about. As I see it, this precariousness – *in* nature, affecting our way of relating to it, being connected to it in its current state of degradation, largely human-caused at that – is a crucial part of *what is being denied* in present-day denial. We want to go on living *as if* nature continues to be what it always was – there for us, safe and sufficient for our needs, emotional and psychological ones as well as material and physical ones – even though we know, increasingly from our own first-hand experience, that nature is no longer like that.

Disappointment is an important part of the picture painted by Lertzman, expressed as a lack of comprehension of how things were allowed to degrade and degenerate so badly, on our watch, as it were. It is impossible to say this without thinking about Greta Thunberg's "How dare you?" directed at the betrayal of adults who have knowingly, and in a sense also willingly, spoiled the world left to their children. The question prompted is as obvious as it is discomforting: "Did it really have to come to this?" – be it the loss of snow in Voss or the pollution of the river in Green Bay; the river that used to be a place of joy and happiness and nothing else; the river, that is to say, that used to be the best place to go in times of trouble, that would always, like other "good objects" in nature, be there as a guardian or comfort but that now has turned into the opposite: no longer a source of comfort but itself a source/object of discomfort, of sadness because ruined in its original, pristine qualities, leaving so many people with a chilling feeling of being "homeless at home", to allude to Glenn Albrecht's term (2019). With the disappointment that this should come to pass questions arise, questions about whose responsibility it is that things have gone not from bad to worse, but from unambiguously good to *having that goodness spoiled* so as to become sources of anxiety and loss instead of enjoyment and happiness; of bad instead of good, of death instead of life. How to face a loss of such magnitude? How to do so without anger and a sense of injustice, of undeserved deprivation – whereby the "undeserved" is problematized by the fact of complicity, of being implicated in the lifestyle and cultural patterns that sustain the deprivation.

Adaptation is key to the answer. "I still love skiing!"; "It's still water, and I still love it!" People do their best to put on a brave face, convincing themselves that it's not that bad, or not all bad. Surely not everything is lost and, besides, one should get better at enjoying what is still, despite everything, left of the good object, the natural environment that was always near and dear.

Lertzman speaks of this as "adjusting one's expectations", that is to say, lowering them so as to learn to live with disappointment (2015: 173). What, more exactly, is this disappointment? It is, as Lertzman suggests, "a softening, a blunter edge than anger or outrage"; it has "a sadder quality, muted, and is entirely livable". As Sally, one of her respondents says, she can live with disappointment since "one adjusts" (176).

What we encounter in individuals whose life quality is clearly – undeniably – negatively affected by ongoing degradation of their natural environment, then, is a never-ending and as such psychically exhausting negotiation between awareness, shock, and disgust and the socially available modes of response or action, to echo Norgaard. Culture, Lertzman helps us realize, enters into this in that it *blocks the types of communication and action that would be effective in addressing the problem*, and in promoting efforts to get at its roots – located as they are in the deep structures, politically and economically, of the growth- and consumption-fixated societies we are part of. There is the ambivalence of being simultaneously aware and unaware of what is going on in terms of deterioration, of losses that are all too real and that multiply and grow, losses that cannot be fully acknowledged to oneself or shared in their depth and gravity with others *precisely because that originate so structurally and deeply*, and are so real, so serious, and so generating of anxiety and despair.

Green Bay in Lertzman's study, like Voss in Norgaard's, are two cases to illustrate what Lertzman calls "a contested site of contradictions": good and bad at once, home to a dominant industry that provides both jobs, securing the good life one strives for and deserves, and despoliation, undermining what one hopes to achieve. Since the good and the bad are so intimately intertwined, albeit in a destructive rather than mutually affirmative manner, Lertzman comments that it is hardly surprising that the concerted efforts of environmental advocates and scientists in the region to challenge the industry, are "less than enthusiastic". The explanation for this goes deeper than the fact that some – a minority – of the inhabitants would risk losing their jobs if the industry were to close. In psychoanalytic terms, what is striking in the response – or non-response – to efforts to tackle the ecological degradation in a meaningful way is that "something is arresting or impeding reparative energies", what Donald Winnicott (1963) conceives of as "concern and creativity". This leads Lertzman to an important observation that can be regarded as her distancing herself from the kind of perspective we witnessed in Stoknes above: "This", she says,

> is ordinarily analyzed as a 'gap' between what people say and do. Again, I reject this notion of the gap because, in fact, the particular space between what people say and do is fraught with psychosocial complications and constraints. It is not a gap at all but a tangle of often competing desires, drives, commitments and investments.
>
> (2015: 126)

Common to the experiences of people in Voss and Green Bay is the hard work of realizing that something "good" and healthy has changed into something "bad" and dangerous – namely the natural environment that provides the framework for their lives. How to tackle this change? In Melanie Klein's psychoanalytic theory, the task of facing such a change and – crucially – one's own part in it raises fundamental issues about the capacity for destructiveness as well as for reparation, understood as making good what one helped destroy: wanting to repair the damage done, to restore the goodness of the object (Klein 1988, 1989). What are the prospects for reparation in a case like Green Bay?

Reparation, says Lertzman, is "not absent but rather lacking a 'home' for broader expression". Why? Because the industrial presence in Green Bay "constitutes an object that cannot withstand anger or redress without arousing intolerable levels of guilt in community members. The refrain, 'Green Bay is a good place to live' is true but only partially so" (2015: 127); its goodness has been spoiled, tainted, it isn't all gone but it isn't what it used to be; the water that used to be clear and safe for drinking and swimming is so no longer. This leads to the situation we first saw Norgaard describe in Voss: "The loss and disappointment, poignantly expressed by the participants, is not a loss that can be socially processed and recognized, which is essential for the work of mourning" (127). The "working through" of mourning (Freud) cannot come to pass, partly because there is no "home" for it, no shared place for it to be accomplished, and partly because the admission of complicity, of playing a part in spoiling the good object, would arouse "intolerable levels of guilt" – a sense of guilt that is no easier to admit, process, and communicate for being *our* guilt, that of us as community members, rather than – be it exclusively – *my* individual guilt. Contrary to what one might have expected, then, the fact that the individual is *not alone* in contributing to the damage caused, and so in the task of facing guilt and the need for reparation, makes the whole process that would lead to guilt being admitted and reparation undertaken that much harder, not easier. The complicity is broader, socially and culturally speaking, and runs deeper, psychically and emotionally, than complicity of a sort that pertains only to a distinct individual, the one "doer" responsible. This is an important insight, though implicit rather than expressed as such by Lertzman.

2.4 Sally's story

A long section of Lertzman's study is devoted to her in-depth interviews with one participant in Green Bay, called Sally. Sally's responses to questions about how she copes with the "industrial ambivalence" that goes with living, and going on living, in a place where environmental degradation is literally occurring before one's eyes are extremely rich, illustrating in first-person manner the complexity of the defence mechanisms in play.

Sally is not born in Green Bay but close to Lake Michigan, making the move with her parents when she was 15. One of the first things she noticed, making for a big change, was that the Bay was so polluted that she couldn't go swimming there; and she had loved swimming in Lake Michigan as a girl. The "shock" of the polluted water, she says, raised her awareness of environmental issues. She goes on to tell Lertzman how her lifestyle has been "filled up" with a physically demanding full-time job, music and adult education. Feeling exhausted and a bit out of breath just listening to Sally's story about how busy her schedule is, Lertzman writes:

> I am most struck by the way she consistently gets close to a topic clearly of great affective investment and then retreats or backs away into a mantra or litany that repeats itself through the interviews: 'There is nothing I can do'.
>
> (2015: 16)

Two things are striking at this point: first, the "topic" referred to is the environmental degradation of her home place, Green Bay, being evaded rather than directly named and confronted, the colloquial "elephant in the room"; and second, Sally from early on makes it her mantra that "there is nothing I can do", before Lertzman has raised this question at all, and so revealing a need on Sally's part to engage in pre-emptive defence, as it were, perhaps seeking – consciously or not – to prevent the issue of her responsibility from being raised.

Lertzman makes much of the fact that in Sally's case, moving house at the age of 15 divided her life so far into two distinct phases: childhood as spent in Michigan, and adulthood as spent in Green Bay. Articulating how as a young girl she enjoyed the water and the sand when swimming in Lake Michigan, Sally comments: "Being young and knowing, not knowing that there's anything wrong, you know the world was a safe place". This, then, describes Sally's world prior to adult concerns, before they would kick in after the move to Green Bay. The move turned out to signal "an end to innocence – from 'idyll' to abject, from the omnipotence of childhood to the imperfections of adulthood, from safety to risk" (2015: 167). The contrast between childhood and adulthood, then, is reflected in the contrast between the safety associated with swimming in Lake Michigan, with nothing to fear or worry about, and the risk, together with a lack of protection, associated with not being able to swim in the polluted water in Green Bay. Sally had to accept, at some level at least, that she had moved to a place that is clearly not safe and "where there are things that *are* 'wrong'" (164), this being Lertzman's way of putting it, not Sally's.

Evidently, to digest the sense in which the change in environment was one from feeling safe to having to be cautious because of the dangers of pollution, came with a sense of disappointment for Sally. Critical questions like "How could this pollution be allowed to happen?" must have crossed her mind. But how, precisely, is this question, with its emotional baggage of discomfort and, perhaps, a dawning sense of anger, to be posed; and to whom? Lertzman cannot but note Sally's evasiveness at this point, observing that "there seems to be a lack of comprehension as to how it was allowed to degrade and degenerate so badly" (2015: 168). Though somewhat under communicated in Lertzman, I interpret this lack of comprehension as a defence on Sally's part against facing head-on a sense that "allowing" things to degenerate – the object going from good to bad, clean to polluted – is a matter of being let down – let down by those who should have known better, those who had the power to prevent what happened, and so to protect her from being exposed to the adversity caused.

Who are "those who" in this formulation? Politically speaking, the answer is probably, most sensibly, the authorities, the experts in charge of safeguarding people and their environment. Psychologically speaking, which is more to the point here, the answer is grown-ups, or the parents. To face being let down like this is clearly something Sally, both as a grown-up today and as a 15-year-old child at the time, does her best to avoid. It is not only the fact that the water has been polluted that is painful and as such calls for a psychological defence. It is also the fact that those who should have provided protection did not do so, making for an act of *betrayal*, to use a term important to my purposes, but not used by Sally or by Lertzman.

In Sally's view, the transition from childhood to adulthood, which as we have seen corresponds with that from Lake Michigan to Green Bay, made it important for her to be adaptable. This is something she comes back to repeatedly in her interviews, referring to things big and small, trivial and fundamental: "You do adapt, because we are, everybody is adaptable" (2015: 172). Having moved, she was forced, as a fact of life, to reorient herself, to become more realistic, to "grow up" with all its connotations. As we saw above, adaptation in Sally's case according to Lertzman takes the form of adjusting her expectations, meaning to lower them, as part of "learning to live with disappointments". Disappointment is a strong thematic undercurrent in the interviews Lertzman conducted; it is a shared experience in virtually all of them, regardless of what their childhood was like, the quality of their relationships with their parents, and so on. To recognize its commonality across so many different biographies is to suggest that disappointment in the sense relevant here comes to the individuals from without, as it were, emanating from their environment in a wide sense, rather than stemming from within, say, from inner conflicts like those typically theorized by Freud. This being so, disappointment is not something that strikes the average person as a rare thing, as conspicuous, but as perfectly normal, though admittedly regrettable.

Having established its commonality, what are the more precise emotional registers of disappointment? Is it an emotion? Interpreting Sally's response, Lertzman describes it as "a softening, a blunter edge than anger or outrage, for example. It has a sadder quality, muted, and is entirely livable. We can live with disappointment; one adjusts, as Sally often says" (2015: 176), whereby her saying "one" rather than "I" signals that the adjustment holds for everybody, not only for her.

Making this move from answering questions on her own behalf to answering them in an impersonal way where everybody is presumably included is particularly revealing when the topic raised concerns agency and responsibility. In several instances, Sally engages in the displacement from "I" to the impersonal people, shifting agency from the one to the other especially when questions about causality come up – for example, what exactly caused the water to be polluted? To ask about this is of course, per implication, to ask about who were responsible, in recognition of the fact that what caused the pollution is not here first and foremost a technical issue, but a *moral* one, directed at those responsible for preventing human-made pollution from occurring. Again, Lertzman doesn't use these terms, this distinction, but invoking it is important to appreciate her point – which is that Sally, time and time again, across various instances, insists that "you can't really prove" the causal link, be it what caused her mother's cancer, be it in what caused the pollution of the water in Green Bay.

We approach here an answer to a question raised above: who to address in cases of disappointment? To the extent that specific feelings are admitted, difficult and negative such, at whom precisely should they be directed? In social studies, the answer often takes the form of pointing the finger at scapegoats, or – alternatively – blaming the victim. None of these responses are forthcoming here. Instead, by questioning the causal links necessary for agency, hence responsibility, hence blameworthiness, to be established in the concrete case (mother's cancer, the polluted water), Sally engages in the art of evasion, appearing to be more interested in letting the people or agencies involved off the hook than in identifying them and singling them out for blame. Why this evasiveness? More to the point, why this evasiveness in cases where Sally herself is clearly not to blame, whereas others – with the requisite power and competence – are? Why see to it that agency – responsibility, blameworthiness – *not* stick to anybody in particular?

Lertzman is evidently puzzled by Sally's response in this case, by her commitment to question the causal links that would allow for criticism, and so – presumably – to get rid of her disappointment, at least until the next scandal of one sort or another crops up. Yet, on my reading, Sally is not really interested in getting rid of her disappointment, presupposing as it – arguably – does that a life *without disappointment*, without reasons to be disappointed, would be possible and so worth aiming for. It is not. Why? Because such an expectation from life, life as an adult having made many sorts of experience, good and bad, is not realistic. It would involve too high expectations. Better, then, to lower them, to be prepared for what happens, since "shit happens"; better, that is, to not risk being taken aback, losing a faith or confidence in life that must prove mistaken, if not in children, then in adults.

When asked directly by Lertzman if the bad things Sally has experienced in life has made her feel anxious, she says no, adding: "You have to adapt to whatever you have. If the water is bad you have to adapt to it, that's just the way it is" (2015: 179). Sally knows that Lertzman is interested in getting her to explore, in some depth, her feelings upon learning about water pollution in Green Bay, yet she responds by moving away from the topic rather than dwelling upon it. At one point she says that "every body of water has their own parasites", referring to a trip she made to Egypt, where she was told not to put her hand into the water. The parasites, she explains, will enter into your skin and "you know it can cause havoc and you can die from it" (179). Why does she mention this when asked about the water where she lives? In effect, what Sally does is to describe water as "naturally toxic", meaning you should stay away from it, not drink it or swim in it. But of course, in the case of Green Bay, the water is not naturally toxic; the very point of learning about Sally's response to it having been polluted is that is has, precisely, been contaminated, spoiled by human activities. Whereas in an objective, factual sense the two cases are very different – the Nile and Green Bay – Sally's eagerness to focus on the Nile reveals something about how she tries, psychologically and emotionally, to relate to the contaminated water where she lives. It is as if the message she tries to convey is "where there is water, there is danger; better take care and stay away".

Again, this is my way of putting it, not Sally's or Lertzman's. Yet it fits well with Lertzman's observation that Sally's paramount message, based on various experiences with water, is the cited one that "you have to adapt to whatever you have". When Sally "seamlessly moves into discussing a body of water thousands of miles away", what she is actually up to is *normalizing the threat* in making this far fetched link (2015: 180). Asked to elaborate, Sally declines, saying she comes to standstill. She then comes back to her "mantra", remarking that "I'm not that concerned about it anymore. You have to acclimate, you have to *accept whatever there is*, otherwise you will make yourself nuts worrying about it" (180; italics in original). Lertzman comments that Sally's demonstrated ability to acclimate and adjust, understood not only as what she herself endeavours to practice, but what she recommends to others as well, is a "strategy of self-preservation". Overall, Sally "seems to have internalized a sense of blame and guilt". In one case after the other, she appears to have developed "an inability to assign anger and aggression towards the source of the problems"; it is as if it has become a lifelong task for Sally to "not express anger or resentment at those responsible" (180), whoever they may be, a task she has committed herself to without having been told by anyone in particular to do so. Protecting others from blame, from being targeted by strong negative emotions like anger, regardless of their responsibility and blameworthiness objectively speaking, is Sally's way of protecting herself in a world where danger, and the discomfort it gives rise to, is perceived as the rule not the exception, as the reality principle of our era. To grow up is to "get real" in the sense of getting used to the world being the way it is, and thereby to give up on the notion – ambition, hope – that it could be otherwise.

Sally's determination to get real by way of adaptation and adjustment comes out most directly in her responses to the question "What can be done?". Whenever the topic of taking action is raised, Lertzman notices a kind of circularity to Sally's narratives, being contaminated as it were by a sense of being entrapped, or stuck. Sally shifts between saying "very little" and "nothing", never explaining what the difference between the two consists in. She mentions her "brief yet intense" involvement with the Save the Whales campaign many years ago, when she was in her 20s or 30s. Lertzman is intrigued by the element of self-mockery that accompanies Sally's story about the rise and fall of her environmental concern regarding the whales. "How will my money help the whales?", Sally exclaims, as if the very idea that there is something effective and meaningful for her to do now, a couple of decades after her involvement, strikes her as naïve, if not downright ridiculous, as proof that she at the time had not yet grown up in the sense described above. In statements characteristic of her way of thinking about this today, Sally begins by saying that "You never want to see an animal go extinct", going on to lament the fact that "You can't eat the fish out there [in the Great Lakes] anymore", immediately adding "I'm concerned about it, I-m aware of it, there's nothing I specifically can do for it" (2015: 185). She then refers to congress and the process of trying to pass legislation to "protect these things", only to repeat her main point: "And there's not really a lot I can do" (185).

Putting Sally's thoughts about her lack of power and agency into perspective, Lertzman observes that the common explanation for why people do not support environmental advocacy (or stops doing so at some point, as Sally did), is "the perceived dichotomy between self-interest and altruistic concern" (2015: 185). True enough, Sally does defend – if that is the word – her fading environmental activism by saying that "retirement is a higher focus to me than saving the whales right now" (184). However, I don't think that the often-cited dichotomy between self-interest and altruistic concern gets to the bottom of a story like Sally's. Why is that? It is partly because the "opposition" between the two is far from compelling – it is well known that when people help others, they not only do the right thing for those others; they report feeling very good about it themselves, also, and the two experiences, or reasons for action, mutually support each other rather than being mutually exclusive, the latter being first and foremost a theoretical construction, though one that seems plausible in present-day neoliberal culture (I shall elaborate this point below). A second and more immediately relevant reason who the "opposition" thesis is wrong in Sally's case is suggested in Lertzman's reference to Winnicott's concept of reparation. Concern, Winnicott maintains, develops when the infant feels herself able to *contribute* (1963: 177). As Lertzman rightly points out, then, "the perception of 'nothing I can do' exists ion a dialectic with the energies mobilized for practice; however, it does not necessarily correspond with the level of concern or care that may be present" (2015: 185).

Before closing this section on Sally, it is worthwhile pausing to reflect on what is at issue here. Let me put it like this: Sally's demonstrated repertoire of defence mechanisms in the face of indubitable, objective knowledge about ongoing environmental deterioration (fish dying in the Great Lakes, water too polluted for drinking and swimming in her hometown) take the form of a psychically straining endeavour to cope with the changes happening by way of adjusting to them qua real. Hence there is no denial as to their reality. Where exactly, then, does denial figure in Sally's story?

On my reading, Sally's story as she tells it is a story about how she has changed as a result of her environment – outer reality – changing. Initially, when first coming to Green Bay and being confronted with its pollution, she was shocked. The pollution was not only something novel and so far unexperienced to her; more to the point, it was *wrong*, something bad not right, contaminated not clean, and thus in need of *reparation* – not in the psychoanalytic sense of Winnicott and Klein, but in a technical and biological sense, involving the task – the obvious task – of seeing to it that what had gone from clean to contaminated be repaired so as to restore it as clean, meaning *as it should be*. What we have seen on the part of Sally and how she has come to change as a matter of the necessity of adaptation and adjustment, making it "liveable" for her to stay in what is objectively a site of ecological degradation, is that she has made the move – seemingly very consciously – from being concerned and showing it to others (by supporting the Save the Whales campaign) to longer being concerned, or at least not as much as she used to be, in her younger days. Why this move? Is it a move justified by, true to, the objective situation? No, rather the opposite, in the sense that the situation has got worse over these years, not better. Objectively speaking, then, there is reason to be more concerned now than 20 years ago. But Sally's change does not correspond to the facts: instead of getting more concerned, she has gotten less. In other words, *Sally's way of changing stands in blatant contradiction to the change in outer reality that she responds to*. Crudely put, the contradiction is one between what is objectively justified and called for (more concern) and what takes place subjectively (less concern).

I am aware that this analysis may seem psychologically simplistic. Can we take Sally on her word when she declares that she is "less concerned" than some years ago? Don't we see her as engaged – trapped – in a never-ending oscillation between attending to reasons to be concerned, and reasons not be, albeit with the twist of asymmetry that she is more receptive to the latter. Indeed, don't we all have a strong, natural inclination to favour a state of mind of not being worried rather than being worried?

To be sure, there is plenty of material in Sally's narrative to confirm the reading that what she is up to is concealing her concern, hiding it from herself as well as from others. Following a popular interpretation of Freud, the work of hiding discomforting feelings and thoughts from myself is a labour undertaken by the unconscious, and so kept at arm's length from what I consciously think and feel. However that may be from an intra-psychic psychoanalytic point of view, my present point is primarily to do with the social dimension, in keeping with we saw in Norgaard's approach above.

Let me explain what I mean by posing the following question: What is being forbidden? What is it that Sally forbids herself to think and feel? The answer is obvious: to be concerned, to admit it, and to show it; to stand for it. In forbidding herself this state of mind, of feeling, of perceiving the environment surrounding her, she conforms to what is expected of her, socially and culturally. And as Lertzman helps us see, Sally internalizes this culturally specific and induced norm, forbidding her to admit to herself what she learns cannot be admitted – expressed – to others: not only how concerned she is, or should we say *in fact* is, despite herself, despite her efforts to the contrary, but also the costs it carries for her to feel compelled to hide what is forbidden from being displayed, the psychic and emotional costs of having to maintain this hiding, day in, day out, with no end, no resolution or relief, in sight.

More precisely, what is the role of culture in the process depicted? To ask what is forbidden for an individual like Sally, is to ask what the culture she is part of commands her to hide – you obey the order to forbid by engaging in hiding it. What does "it" refer to here? On an overt plane, the fact of environmental degradation Sally is a first-person witness to. On a deeper plane, the psychological one, what Sally is culturally obliged to hide is not something "out there", facts about objective changes in the outer world; rather, what she is obliged to hide is *part of herself*, understood as the repertoire of sensations, feelings and thoughts, as well as needs and drives, that constitutes her as a singular person, meaning her*self* as an authentic self. In effect, then, culture forces Sally to abandon her authentic self, or at least large portions of its repertoire, compelling her to engage in what it takes to put into place a *false* one in its stead – "false self" being Winnicott's term for the sort of self that is created by way of adaptation and adjustment to a social environment that sanctions manifestations of the authentic self; sanctions, that is, the sorts of feelings and thoughts that are culturally considered as "unwanted", or "unhelpful", or simply "wrong", be it for moral reasons.

In this sense, then, the culture that Sally has to conform to demands that she internalizes the "correct" way of discriminating between what can be felt, thought and – thereupon – expressed, and what cannot. In Sally's case, being concerned with the fact that the environment is being degraded because of human-caused activities is what cannot be expressed, not in the free, spontaneous and psychically–emotionally rich sense of Winnicott's authentic self. What we see, then, is that Sally's culture compels her to adopt a false self. Over time (and I shall expand on this below), this false self, one that forbids itself unpleasant feelings and discomforting thoughts about what happens to the environment, becomes the socially shared self, the default position for selves in the present era, in Sally's community and countless others in Western industrialized countries affected by serious environmental degradation.

2.5 Adaptation as a case of perversion

It is worthwhile dwelling on this point, and ask: Is the way culture imposes a false self on an individual like Sally, a case of perversion?

It may seem far-fetched to invoke the notion of perversion in this context, and to be sure, it is not a concept employed by Lertzman, let alone by Norgaard and Stoknes. However, novel phenomena call for novel concepts, and for the imaginative stretching of old ones to test their potential validity in cases other than those for which they were originally intended.

In Susan Long's study *The Perverse Organisation and its Deadly Sins*, mentioned in the previous chapter, various case studies highlight what Long calls the "destructive nature of collective perversity", the downfall of the company Enron during the financial crisis in 2008 being a case in point. The corporate sins Long concentrates on are pride, greed, envy, sloth and wrath or rage. Her claim is not that all members of the companies she studies demonstrate these traits in their behaviour, or that the bosses at the top demand that all employees do so. Long's perspective being that of a critique of power, that is to say, the asymmetry of power between those at the top and the rest, her point is about how each member of the company is forced to get used to the collective perversity, to tolerate it as something "necessary", even imperative for the company to survive in a world of dog-eat-dog competition. In particular, employees are expected to tolerate "the domination of narcissistic pleasure over the common good; the concurrent denial and acknowledgement of reality in the turning of a blind eye, and the development of perverse certainty" (2008: 150). At times, the corporation displays tendencies associated with individuals, such as neuroses and psychoses, as well as splitting, projection, denial and negation, a similarity that occurs because "both individual and organisation share the fundamentals of human system dynamics that are located in culture" (153). In line with my analysis above, "culture" is not meant in a general sense here; rather, this is about the culture characteristic of a society where market relations make inroads on a number of arenas and roles, rendering these more instrumental and geared to personal advantage in a situation of competition, and where individuals from early on are encouraged to see themselves as consumers and customers rather than citizens, in a time where people lose faith in their governments. People lose faith and become disillusioned by witnessing time and time again how their government and elected political leaders – hence not only corporate CEOs – allow both individual greed and corporate greed to take primacy over the common good they are, or until recently were, supposed to safeguard.

But why is perversion a helpful notion to describe this? Isn't the word too strong, too tied up with something scandalous, referring to an individual's deviation from the norm?

My answer is as follows: The perversion I identify is to do with something truly scandalous becoming the new normal, the "reality" you have to get used to. And in the cultural pressure to get used to it, it doesn't suffice to turn a blind eye (Long) to what is happening around you. Taking a step further, you are required to engage in denial about the consequences, either by flatly denying they are real, or in embracing the turn-around by which *what used to be vices are made into virtues*, with greed enjoying pride of place. The turn-around is an act of *reversal.* That greed is the new "good" captures the essence of perversion in a cultural context: it is not a matter of a single individual's convincing himself that what used to be bad is in fact good, so that there is no reason for shame, for hiding it away; rather, it is a matter of the individual's culture demanding him or her to be complicit in the reversal of virtues and vices.

Given this clarification, the term "perversion" begins to seem less far-fetched than when introduced above. Consider the material we have looked at so far: Norgaard speaks about how denial is thoroughly "socially organized" rather than a piece of individual behaviour; Sally narrates to Lertzman how she tries to adapt and adjust to the new normal of Green Bay. Common to Bygdaby's residents and to Sally is their having to comply with a cultural demand to avoid using the word "bad" about changes that objectively speaking are all about degradation and deterioration, and thus ample cause for being anxious and wanting to take action, action of a sort and magnitude that needs to be communal, structural, and political to be up to the task. Eirik, a Bygdaby resident Norgaard talked to, puts it like this: "There were many things that we felt that we should have done, but we did like the others anyway, right? Because you have to fit into normal society. It is difficult. […] If you did everything entirely ideal, you would be an outsider in the society"

(2011: 97). Speaking out – truthfully, authentically, and spontaneously – about the perceived "badness" is precisely what is being forbidden, forcing a person like Sally's emotional reactions and troubled thoughts into hiding, in effect turning them into something shameful, something others should not see, and would condemn or ridicule should they notice despite the recurrent work of hiding them whose strongest tool is that of denying them to oneself.

Perversion here takes the shape of pretending that what is bad is good; that one should be, and expect others to be, positive about something that is deeply disturbing. Or put differently, *of course* one still believes in "progress" and that things will turn out for the better, so that the changes occurring will be "worth it". Perversion in this situation, then, does not *directly* assume the form described in Long's study, where vices are turned into virtues. But nonetheless, it a case in point of a similar exercise in reversal.

Given this perspective, let us take a final look at Sally.

There is a tendency in Letzman's account to be more attentive to Sally's way of talking than to her way of acting. In a sense, this asymmetry is inevitable in that Letzman's main access to Sally's behaviour and how it has changed over the years is Sally's way of talking about it. That said, what I wish to highlight more than Letzman tends to do is the many subtle ways in which Sally has changed her everyday habits. For example, at one point she mentions that some years ago she stopped talking a walk from her home to a nearby place which offers a good view of the Green Bay area. She liked going to this place to enjoy the view of the water, allowing her to see the places where she used to swim, bringing back fond memories of the time she spent with family and friends there on Sundays. After she had to quit swimming because of the pollution, however, she stopped seeking this view. Why? For the same reasons that are familiar to people experiencing "ecogrief" in all sorts of places around the world these days: seeing the water would inevitably, involuntarily, engage her in an exercise of comparison, bringing back memories of how the water used to be – clear and clean – in contrast to how it is today – dirty and polluted. To be drawn into this comparison is not something neutral; nor is it something purely cognitive, as a simple act of registering the contrasting view between now and then. Instead, the change is emotionally charged, prompting feelings of anxiety, sadness, and loss, since something associated with joy and positive memories had been ripped away, as it were, leaving a site of deterioration, of *a good object being spoiled and turned into a bad one*. To process this change emotionally is to face vulnerability, the cost of being exposed to the significance that something used to hold, and that it holds no longer, which means exposing oneself to bereavement. Better, then, not to expose oneself to being reminded of this change, namely be staying away from the place with access to the view. Sally physically avoids the place with the view to avoid the psychic pain – trauma – that it carries as its new meaning.

What I am highlighting in Sally is not something Lertzman completely ignores or fails to acknowledge, but rather an aspect of Sally's endeavour to adapt that tends to be overshadowed by the spoken – or unspoken – word. To be sure, Lertzman is right to point out the many instances where Sally is *evasive* in her manner of talking. Yet, and this is my point, what Sally *avoids* in her way of acting, playing out physically in the example about not going where she used to go, is no less significant, and no less subtle as a manifestation of ongoing adjustment to the new normal. To Lertzman, then, Sally's way of responding to her questions goes to describe "how her daily practices and lifestyle serve as effective "defences" against experiencing certain affects and anxieties", observing that "she is so exhausted she literally doesn't have time or energy to contemplate these issues", that is, those of environmental degradation in her home area (2015: 191). Sally, in a way typical of so many people these days, throws herself into her work and her hobbies, such as music, making sure that hers is always a very busy schedule, with days filled to the brink with things to attend to, craving all her cognitive and affective energy. Though important in their own right, in their abundance these activities total as so many distractions, keeping at a – hopefully – safe distance the causes of her discomfort, making the – conscious or unconscious – decision to avoid the places that would hurt most a useful strategy. In line with this interpretation, Lertzman concludes her reflections on Sally by noting that the move to Green Bay "registered a loss on such a scale that it led Sally to withdraw into a smaller world and to accept a level of resignation and "adaptation" as a way of life" (194). Echoing what we saw in Norgaard's study in Bygdaby, the lesson is that "if loss does not find a "home" to process, for example, social forms of acknowledgement or support, it may go underground or become expressed in more defensive ways, such as filling up space to avoid any sense of loss or impotence" (194).

2.6 The nature of apathy

This leads me to a final issue that Lertzman helps us appreciate, namely the nature of apathy.

Apathy is a well-established topic in the literature on climate psychology, holding a key place among the responses – or non-responses – that are so richly present in the surveys, thus demanding an explanation. Norgaard points out that apathy carries "a connotation of deficient moral qualities such as laziness or greed that prevent people from engagement" (2011: 58). She finds this misleading, and instead proposes that the apathy observed with respect to environmental problems not be seen as lack of caring, but as "the refusal or inability to experience pain". Thus understood, apathy is closely allied to denial at its root. Begging to differ from the dominant social movement perspective that non-participation is the norm (itself as sign of disillusion, easily recognizable on the part of many activists, I add), Norgaard maintains that "apathy takes work to produce", citing the scholar Nina Eliasoph: "Simple apathy never explained the political silence I heard. Instead of "apathy" was a whole underwater world of denials, omissions, evasions, things forgotten, skirted, avoided, and suppressed – a world as varied and colourful as a tropical sea bed" (2011: 59). Stanley Cohen makes the same point: "We can feel and care intensely, yet remain silent" (2001: 9). People do care, Norgaard insists; "they just aren't able to express their caring in all social contexts" (2011: 59). Using the term apathy synonymously with non-action and non-response, she stresses that the lacking response – mobilization – so often lamented in discussions about climate change is not "an issue of lack of information or concern"; rather, it is a result of socially organized denial, in keeping with Norgaard's overall analysis. Her claim, then, is that apathy is not a manifestation of something that is lacking, conspicuously so, but instead attesting to something that is present, only that it cannot be expressed as such, but has to be suppressed and hidden, much like we saw with Sally. The fact is that "information on climate change [is] *too disturbing* to be fully absorbed or integrated into daily life", so that the alliance between denial and apathy must be understood as "testament to our human capacity for empathy, compassion, and an underlying sense of moral imperative to respond, even as we fail to do so" (2011: 61). Given this understanding, "apathy is the mask of suffering" (59).

Norgaard's analysis agrees with Lertzman's in that both regard apathy as a thoroughly *reactive* phenomenon, as opposed to a first mover, as it were. Hence apathy must always be understood in light of what it is a response to. In part, this is why approaches that begin with the premise that apathy is about non-response or non-reaction set out from the wrong end. Indeed, when we discussed Sally we saw that the behaviour that we, on one level, may describe as apathy in the sense of "not acting", or having stopped acting, in the face of environmental degradation, on another level takes the form of fullness rather than lack, presence rather than absence, borne out in how Sally came to cope by filling up her daily life with "all sorts" of things to do, except things that would confront her directly with the crisis at hand. Sally at one time tried this, by joining the Save the Whales campaign, only later to retreat from that involvement, the one of overt action of a sort that testifies to her recognition of the problem and the urgency of it being tackled. In Sally's case, then, the lack of engagement in action that Lertzman reflects upon by way of summing up her entire trajectory, is motivated twice over, as it were: it is not only that she has a history of experiencing first-hand the change from good to bad object (the water in Green Bay); she also has a history of actively responding to that change, or to the environmental crisis more generally, by joining a political movement. The apathy-cum-resignation Sally expresses at the time Lertzman concludes her interviews with her results from a double disappointment, not only that stemming from the object turning bad, but that stemming from experiencing that attempts to do something about it turned out to be ineffective, if not downright futile. As if this were not enough, Sally also observes that since she withdrew from activism, things have gone from bad to worse, reinforcing the lesson that "nothing can be done" and justifying her feeling of powerlessness.

A question worth asking here is this: Does Sally exhibit the sort of "inner conflict" that so many climate psychologists are preoccupied with, Stoknes among them? Or do we encounter something else in her case?

A typical way in which such inner conflict is exhibited, is in the form of *hypocrisy*: a person declares commitment to one set of values, only to pursue another – a contradictory one – in her way of acting. "You say you care about the environment, but the fact is that you fly instead of traveling by train, again and again". Is Sally a hypocrite in this sense?

I would say no, and I think that the reasons why are worth spelling out in a way not found in Norgaard and Lertzman.

In my view, Sally's case is especially intriguing for the following reason in addition to the ones already examined: over the years, she has learned to undertake the tough psychological work of downplaying, to the point of obliterating, the part of her that would correspond to the "I declare that I care about the environment", replacing it almost entirely with the part corresponding to her not caring about it. That she takes it upon herself to hide, to the point of extinguishing, the caring part, recalls my suggestion above about perversion, understood as the individual's obedience to a cultural imperative to adapt to a situation where something bad spoils something good, and to pretend that this adaptation is perfectly normal and not to be regretted or fought as such. What is noteworthy in Sally's case, as distinct from Norgaard's material, is how modest a role *interaction with other people* seems to play: Sally's way of coping with adverse change is almost entirely located in her relationship with herself, rather than in her relationship with others, be it her co-residents – neighbours, friends, colleagues – in Green Bay. The other-directed, genuinely *social* dimension of "organized denial" in Norgaard's optics is not totally absent in Sally's case; it is presupposed as having some impact, to be sure. But its role is largely a tacit one, secondary to how she relates to herself and tries to maintain some kind of inner balance or harmony, to recall the motif I mentioned earlier from Socrates when discussing Gardiner in Chapter 1: the primacy of the self-relationship.

Perhaps the explanation for this is the simple one that Sally is single: she lives alone. And indeed loneliness strikes the reader – at least me – as an undercurrent in her life, though not as such brought up by Lertzman, it seems – be it for not wanting to be intrusive. That said, the mechanisms of internalizing the social milieu's and the culture's norms of what to feel and what thoughts to share with others are operative regardless of how much – or little – Sally is socializing with others. The fact remains that the part of her that "cares" is being so downplayed and so silenced over the years that it is shifted – or to put in the active rather than passive form: is shifted by Sally – from being "front-stage" in Erving Goffman's (1959) sense to being "backstage"; from being what she identifies with and stands for, at least to a large extent, to what she disavows and dissociates herself from.

One could say that one of the rewards of this shifting is that there is no hypocrisy: the inner conflict or tension that so preoccupies scholars like Stoknes, is virtually absent in Sally, since she has successfully made the transition described so that if there was once a gap between what she said and what she did, or between what she truly felt and what she communicated to others, this gap has been closed.

To me, there is something very troubling, even scary, about this ability to close the gap, to erase the tension. I think it raises critical questions about Norgaard's, and to some extent Lertzman's, conclusions about apathy, holding that apathy, and the way it colludes with denial, is "testament to our human capacity for empathy, compassion, and an underlying sense of moral imperative to respond, even as we fail to do so" (Norgaard 2011: 61). On the one hand, I agree that the apathy pointed to in individuals, and often explained in terms of cognitive dissonance a la Stoknes (knowing but being unable to act on that knowledge), needs to be de-individualized and to be seen not as a more or less conspicuous absence of engagement and lack of action, a deficit on the part of the individual, but as a reactive defence mechanism upheld by culture. In a word, what culture helps these individuals do, is not to tackle the environmental crisis in an effective way, but to make sure that they avoid its gravity and urgency. On the other hand, however, I am not convinced that a case like Sally's serves to confirm the validity of Norgaard's quite from Cohen, that "we can feel and care intensely, yet remain silent" (2001: 9). To support her claim, Norgaard refers to studies of empathy and prosocial behaviour that find that "both people who experience very low levels of empathy and people who experience very high levels of empathy are less willing to help victims in a variety of situations", picking out as particularly relevant the finding that "people with high levels of empathy for victims found encounters so painful that they avoided involvement rather than assisting someone in need" (2011: 59).

The problem I see with this reasoning is that the empirical evidence cited is unsuited to throw light upon Norgaard's claim with respect to people's way of responding to climate change. There simply is no parallel between the readiness, based on a person's empathy, to assist others in pain, and the stance taken towards the adverse effects of climate change. There are no victims crying out for help in a sense calling for attending to their acute pain in the context of Sally's story. Indeed, a person's "level of empathy" as referred to by Norgaard is ill-suited to explain the reasons why people respond differently to how "their" environment is affected. Empathy in the sense presupposed has the conventional form of a person's reaching out to human others, perceived to be in distress and in pain (see Vetlesen 1994; Camassa 2024). By contrast, climate change affects not only humans like myself but also the natural world – consisting of everything from water to whales (to recall Sally's story) – in a sense that explodes the anthropocentric, towards-fellow-humans-only setting for the display of empathy that the studies Norgaard cites are restricted to.

The deeper question here is whether empathy is the right kind of capacity to be looking for, let alone to "measure" on an individual level, or whether the "caring" that has as its addressee other-than-humans of various sorts – or simply, nature – has a different affective origin, be it a feeling of attachment, or belonging, to a natural, multispecies world whose ongoing destruction is difficult to take in. As we shall explore in more detail below, ecogrief, understood as triggered by a readiness to *do without denial*, to resist it, and so to "take in" the degradation in question, is not first and foremost a response to pain but to *loss*: the sadness of being witness to, and to some extent complicit in, the ongoing human-driven assault on the natural world and its immense richness and diversity, its capacity to flourish in so many species-specific ways.

2.7 For and against alarmism

Having discussed Norgaard's and Lertzman's studies on denial, I now return to the approach pursued by Stoknes.

At one point, Stoknes takes a step back from the many books written about how denial poses a serious obstacle to attempts to tackle climate change, remarking that "denial and self-deception are part of being human", citing the poet T.S. Eliot: "Humankind cannot bear too much reality". Like it or not, the fact is that "we cannot become conscious about every aspect of our lives" (2015: 79). Two pages later, Stoknes resumes his main topic, namely how countless efforts by scientists and others have so far failed – and failed spectacularly – to "break through the wall that separates what we know from what we do and how we live" (81).

We recognize in Stoknes's first comment the kind of humanist assessment of denial that we saw in Norgaard, and to a somewhat lesser degree in Lertzman: denial is human, all-too-human, and no one among us would be able to live without resorting to the "protection" that denial supposedly provides. Against the backdrop of this sweeping generalization about human psychology, Stoknes reminds his readers of the failure to address climate change, calling it "the greatest science communication failure in history". The lesson we have to stomach, then, reads "the more facts, the less concern" (81).

That the so-called "information deficit model" has proven wrong is a well-established fact and need not concern us now. More important at this juncture is the proposal Stoknes offers to get out of the current impasse. We need, he says, to tell better climate stories; in particular, we need to "avoid apocalypse narratives, and instead tell stories about green growth, happiness and the good life, stewardship and ethics, and re-wilding and ecological restoration" (2015: 149). The urgency of switching to a more positive, upbeat storytelling about climate change is borne out by "the dominance of the disaster narrative in more than 80 percent of news", making it a major task to overcome the fear, guilt, and helplessness sustained by "apocalypse narratives", and so helping us, at long last, to overcome "our deepest barriers, particularly denial and identity" (149).

Stoknes's assertion that "apocalyptic" story-telling and alarmism will prove counter-productive, prompting feelings of fear and helplessness rather than mobilizing people for political change, plays a considerable role in his argument. But is it borne out in empirical studies and in what we have witnessed the last years? I don't think so. Take Greta Thunberg. What she did was to *disobey* Stoknes's advice, consistently and unapologetically: She sounded the big alarm; she painted in scary dark colours the future the proverbial "grownups in the room" are sleepwalking us into by their inaction and foot-dragging in things political-environmental – and succeeded in mobilizing tens of millions of young people world-wide to take to the streets in 2019, in the biggest political movement ever seen in history for that group. True, not everybody responded by being mobilized; many, especially 60-plus men in countries like Norway, reacted aggressively, calling the 16-year-old Swedish girl all sorts of things; and one doesn't have to be a psychoanalyst to find their excessive aggressiveness highly revealing: evidently the girl hitting the alarm button had hit some sore spots in their psyche, their sense of entitlement in particular. But then again, people are different, with different ways of responding. If anything, Thunberg's "alarmist" messaging was clearly contrary to what Stoknes considers a wise thing to do, proving – at least for a period – a resounding success.

Moreover, I am not sure that Stoknes is correct that "more than 80 percent of news" is of the apocalyptic kind, or one that deserves the label. This is definitely not the case in Norwegian media. Besides, not all the news is about climate change. Most is about all sorts of other issues, and to the extent that "environmental problems" occur – be it in the form of typhoons, or floods, or draught – their sheer frequency these days tend to "normalize" them and so marginalize them: they are exactly *not* news but the new normal, which is something else. Yet my main point is not about media statistics, but about how to understand a person like Sally, as well as the residents of Norgaard's Bygdaby.

Regrettably, there is no Sally, or Eirik, in Stoknes's book; but there is his own personal story, that is, the parts of it he wants to share with his readers to underscore his points. He is content to advance his anti-apocalyptic proposal of storytelling on a most general level, such as we saw in his reminding us of our need to protect ourselves against discomforting information by way of denial, confirmed as it were by what we know, if we are honest, about human nature. The same sweeping style holds for Stoknes's repeated remarks about humans being "short-sighted", a notion he takes over, uncritically, from his collaborator and mentor Jørgen Randers.

If we look at the interview-based material in Norgaard and Lertzman in light of what Stoknes is saying here, what do we find?

Let me concentrate on the findings discussed above concerning apathy. The apathy met upon in the persons interviewed in both studies is theorized as the endpoint of a long process. A common theme is how efforts to demand that necessary changes – of policy, of economics; of lifestyle – take place are given up following a series of what Lertzman calls "disappointments". In this regard, it makes no real difference, in people's perception as well as experience, whether the measures hoped for are to be carried out by those at the top – say, the level of government – or by local authorities, accessible in the community in the form of politicians or experts known through one's social network. The lesson is that whatever the particular level and types of agents (authorities) called upon to take action, and regarded as *responsible* for doing so, the results are the same: nothing much happens. With time, withdrawal from engagement, retreating from activism, shifting attention and energy to other activities – more upbeat ones – such as music or joining a choir is the common response, and as such perfectly understandable.

But isn't this way of putting the findings is fully compatible with Stoknes's? So what is the problem?

The problem – or Stoknes's – is that the "world out there" that the people in question respond to in the way described, is not a world filled to the brink with narratives about apocalypse, evoking "fear, guilt, and helplessness". In fact the opposite is closer to the findings: the culture in which these individuals are entrenched and the social world they are part of, and wish to remain so, convey the message that, overall and in the long run, "things will work out fine", that there is nothing to be seriously worried about, and in cases where this message is defied, and worries – apocalypse-like or otherwise – are actively shared with others, such attempts at sharing is quickly and massively blocked: they "kill the conversation", to recall Norgaard's observation.

Accordingly, the apathy with which someone like Eirik or Sally is "diagnosed" by the scholars interviewing them is of an altogether different kind than described by Stoknes, adopted as a reactive mechanism for different reasons than those emphasized by him: what results in withdrawal, resignation, and apathy is the insistence on business usual, with its stubborn belief that things will work out; an insistence that surrounds Eirik and Sally and that they are constantly expected to share and remain loyal to, indeed to actively help reproduce even if they are deeply uncomfortable with it, feeling it to be more a part of the problem than the solution, granted that they allow themselves to feel that way at all. Eirik and Sally are not worried because of the (alleged) dominance of apocalyptic narratives, but because of the stronghold of *narratives of the opposite kind*, suggesting "we will make it, and no changes of any magnitude are needed for us to do so". In other words, they are worried about the way in which any narrative, any feelings and thoughts, that testify to *being deeply anxious about the future* and what it holds are being discouraged, silenced, and made to disappear. This is what strikes them as so overwhelming, culturally and socially speaking, that they react by way of helplessness and feelings of powerlessness. That they react this way is a manifestation of the emotional and psychic costs of feeling compelled to *internalize* the cultural silencing at work, underpinning and echoing internally what is communicated socially, with others struggling with the same imperative, yet without there being much comfort, let alone potential for resistance, in the fact that others suffer the same strain as I do (I shall explore this more fully when discussing resistance in Chapter 4).

Being constantly *reassured*, then, told that things will work out fine, is at the root of the psychological responses encountered. If "apocalyptic" narratives are about sounding the alarm with respect to the urgent need to change – structurally, not only in terms of individual behaviour – then people like Eirik and Sally help us see what happens when there is *too little* of such a narrative in the public sphere as well as in the (their) private one, not too much, as suggested by Stoknes. On my reading, then, what is experienced as truly scary is the mismatch between what needs to be done and how little is actually being done. This is not a mismatch, or a conflict, within the individual and his or her psyche, as theories focused on cognitive dissonance have it, but a mismatch occurring and playing itself out, with disastrous consequences, in the real world.

In talking about "what needs to be done and how little is actually being done", the mismatch I refer to is located in the relationship between "those in charge who can make a real difference" (governments and corporations) and the real world where crises not adequately attended to go from bad to worse. It is the failure to take action on this level that is perceived as such, in its seriousness, by a person like Sally. Accordingly, her entire repertoire of feelings – mixed such – are directed at this failure on the "top level" (recall Sally's remarks about congress, about passing laws), feelings spanning anger on the one end of the spectrum and apathy on the other.

In putting it like this, am I in agreement or disagreement with Stoknes's analysis?

It is tempting to answer that of course Stoknes will concur and that there is nothing in his perspective that suggests otherwise. But note how his way of framing the issue differs from mine. In approaching the phenomenon of denial by way of "cognitive dissonance", Stoknes's focus is from the outset "what happens within the individual", in contrast to "what does the individual think and feel about the mismatch, involving those in charge, between what they say and what they do", as my focus has it. To the extent that there is a mismatch, giving rise to conflict and tension, and the feelings that accompany them, it is one located within the single individual in Stoknes's approach, as distinct from the extra-individual, outer world context where I locate it.

Having set out the contrast between the two approaches, the question is: Do they contradict or exclude each other? Or are they compatible?

It is true that a person like Sally may be said to exhibit both at the same time: on the one hand, she *is* concerned about the big picture and the failure of those at the top to take action; and on the other hand, she *is* caught up in the sort of inner conflicts that result from trying to adjust to the uncomfortable reality surrounding her. In this sense, then, Sally can be seen to both exhibit the mismatch that I am describing going on "out there", *and* the conflict between "saying one thing and doing another", or "knowing and not knowing at the same time" that Stoknes is preoccupied with. There is nothing, then, that compels us to conclude that the one sort of concern on Sally's part must exclude the other; they can perfectly well co-exist – co-habitate – in her life as she has come to live it.

The deeper question, though, is whether this co-existence of two dimensions of mismatch or conflict – one located outside, the other within one's psyche – that may well, on paper and prima facie, be perfectly possible, is actually borne out in practice. There cannot be a general answer to this, one that would apply to all individuals. That said, in Sally's case what is striking is not the co-existence (simultaneity) of the two, but how they gradually move apart and end up being separated, split. Indeed, this very process is at the heart of Sally's trajectory and the way she tells Lertzman about it: she was "shocked" when she first came to Green Bay and experienced first-hand the pollution there, in her new hometown; gradually afterwards, over many years, she developed strategies to "live with" what initially shocked her, hoping to come over and get rid of the mixed feelings she was suffering from at the, feelings including rage directed at those "on the top". The rage wore off, as it were, as a consequence of the coping strategies she practiced; so too did her initial engagement in the form of activism, of trying to pressure those in charge to get their act together, to enact the responsibility to set things right that goes with their authority and their power. The endpoint of this trajectory is the version of Sally that Lertzman encounters in her interviews: a person who has largely given up hope that things will in fact change, that those in charge will do as they should do; a person who strikes the researcher as having withdrawn into her own world, with its busy schedule, leading a life where environmental activism no longer figures.

Again, I am not claiming that Stoknes would disagree with this portrait. But the fact is that it does differ, in its emphasis, from his approach. Let me put it like this: Stoknes starts his inquiry too late. What he says captures someone like Sally *not* from the initial phase of her trajectory and to its present-day end, covering it in its entirety, both temporally and psychologically speaking, but instead restricted to Sally in her *present* stage and state of mind, theorized as how she copes with her cognitive dissonance. Stoknes misses out on the full dynamics of the process, one involving important changes and shifts, instead zooming in on how it comes to be that conflicting thoughts and conflicting feelings co-exist, in the lived present at that.The predominance given to the individual, with a keen eye for his or her inner conflicts, is something Stoknes shares with Gardiner, whom I criticized on that point in Chapter 1. But, one may well ask, does it really make sense to be critical of theories that set out to explore the role played by denial these days, and that do so by way of concentrating on the individual? Isn't denial something that is either present or absent in individuals? Something that persons, like Sally or Eirik, resort to as part of their individual life stories? And something that varies not only within that individual trajectory, but varies between different persons as well?

It is tempting to answer in the affirmative, and to add that it could not be otherwise. But there is nothing in the phenomenon of denial as such that limits its presence, its influence, and extension, to the single individual as such. In being socially organized, as Norgaard in particular seeks to show, denial is a cultural force, profoundly entrenched in the networks and interactions that concrete persons, living in concrete places, are part of. Denial is such a powerful phenomenon precisely in being shared, nothing strictly private but integral to the social fabric, which is why (public and open) attempts to resist its pressure means risking to become an outsider (Eirik) and roundly condemned as being a traitor (as experienced by young climate activists in Norway). Of course, to say this is hardly news at this point. Why then do I find it important?

What I am getting at is the dialectic between culture and social structure, on the one hand, and the individual on the other: the ways in which they interact, mutually influencing each other. A single individual can make a big difference to the culture: Rosa Parks and Rachel Carson did, Greta Thunberg did (or tried to). When a particular individual *shows the way* – as it comes to be called, in retrospect – his or her way of acting, setting an example, helps change the culture; and that changed culture then influences the individuals making it up to change their ways – that's the dialectic. The effects vary, of course: there are the "this changes everything" situations alluded to in

Naomi Klein's (2014) famous book title; there are massive and enduring shifts, like the abolishment of slavery; and there are short-lived ones leaving no lasting mark on history.

This dialectic will be a recurring topic in the rest of the book. For now, the following observations are worth making.

There is an undercurrent in Sally's story about how she has lived through various phases of coping with being a witness to environmental degradation in her home place. It is to do with agency in general, and responsibility in particular. We have looked at the agency bit in great detail when discussing Lertzman. The responsibility bit has been less thematized. How does it play out in Sally's case?

The point I wish to use Sally to make here is about the tendency in many individuals to take some responsibility for the crisis, the mess we're in; and doing so in a situation where those having responsibility fail to take it. I take it in their stead, as it were; I take it upon myself to do so by way of *compensating* for the failure at the top. When exactly is this tendency exhibited in Sally's story? One could say that in volunteering to do work for an environmental campaign (Save the Whales), Sally is doing something that she really shouldn't need to do: the government, together with relevant organizations, are the agents who should take action, in terms of resources, competence, and authority. But Sally, and others like her, see that they fail – and responds by taking action, bottom-up, hoping for the appropriate response on the higher levels of power. Nothing very special about this, one may say; it sounds like the way democracy works, offering ample opportunity for individuals to mobilize themselves as *citizens*, pushing things in the right direction and being concerned with the common good.

This model of democracy is fine, but for Sally and countless other one-time-activists, is doesn't live up to its promise in practice, leading to disillusion. Nonetheless, the inclination for concerned individuals to take the impasse of current climate politics upon themselves in the form of assuming personal responsibility to set things right, be it by pushing those higher up, or by adopting a more sustainable lifestyle, is strong in present-day society. It fits with broader cultural trends not linked to the climate crisis as such; trends encouraging the individual to assume responsibility in evermore areas of life, "to be the change that you want to see in the world", "to be the best possible version of yourself", and to build resilience to survive in a "harsh and senseless world", to cite Lertzman's depiction of how Sally has come to look at it (2015: 187).

The call for resilience will be a major topic below. I mention it here because it helps shed light on Stoknes's analysis. In my view, he fails to do justice to the dialectic between individual and culture that I find crucial to the prevalence of denial in a society like Norway. To see this, let us put a simple question to Stoknes's account of denial: in denial about what?

The answer is clear: as signalled by the title of his book, the denial Stoknes talks about is at work in the strategies individuals employ in order "not to think about global warming". In its pure form, denial in that sense is exemplified in people who come across as "climate sceptics": they refuse to "believe" in the science that informs us that climate change is real, and that it is (mainly) human-caused.

To be sure, this notion of denial seems to go to its essence: an "inconvenient truth" (Al Gore) is being denied, and the truth in question concerns the very fact of climate change. But there is much more to denial than this, as Norgaard and Lertzman have showed us. The question "In denial about what?" has several answers, and therefore denial comes in several versions. As Sally in particular helps us realize, due to the richness and various phases of her trajectory, a person can be in denial about *what is being done* to tackle the problem, granted it is acknowledged as such; about the *trustworthiness* of those obliged to do so; and about the *efficiency* of measures like technological innovation and entrepreneurship (Elon Musk's Tesla), "ecosystem services" and "decoupling" of economic growth from environmental overshoot, promising to make growth "sustainable". Common to the types of denial listed here, selected by me for present purposes, and different from the typology found in Cohen's classic study, is that they manifest what I call "false optimism": holding on to a belief that things will work out fine, if only the right agents and the right measures are devoted to solving the problem, hence succeeding in averting the crisis.

I alluded to this above when I spoke about *reassurance*. As individuals we want to be reassured, to believe that things will work out fine; and we are part of a culture and society that, in so many subtle and tacit ways, convey and sustain such reassurance, supplying what we want, as it were.

For individuals like Sally and Eirik, there comes a point where the reassurance offered in their small world as well as in the big one no longer appears convincing. Doubt creeps in, critical questions are asked, receiving unpersuasive and evasive answers, killing the conversation. If there is nowhere to go with the resulting worries, they are inwardly suppressed, hidden.

Is denial involved here, or are we stretching the word? It is about losing faith in the dominant story of reassurance, and in that sense denying its truth, refusing to go along with it, to partake of what I call "false optimism". There is no longer any relief in subscribing, like so many others, to the optimism expected; going on doing so is not admirable, it is hypocrisy, joining in a vast exercise of *pretending* that things will work out fine.

2.8 Please end on a note of hope

This brings me to a final look at Stoknes's position.

In his book's last chapter, Stoknes asks: "Why do climate books and speeches have to end on a note of hope?" (2015: 217). This is a pertinent question, and I have met it myself on countless occasions, with all sorts of audiences, from academic ones to 10-year-old pupils. Commenting on the fact that politicians running for office "all say they were born die-hard optimists", Stoknes writes: "But optimism by itself may easily become overblown and spill over into wishful thinking. Insisting on positive thinking, simple solutions, and bright futures may devolve into a type of denial by itself". He goes on to say that "both pure optimism and pure pessimism make poor scenarios" (218). And he cites, approvingly, poet Gary Snyder's reply to the question "Why bother to save the planet?", namely: "Because it's a matter of character and a matter of style" (219). The closest Stoknes comes to answering the question on his own behalf, is this:

> So yes, it's hopeless *and* we're going all-in. The active sceptic gives up the attachment to optimistic hope and simply does what seems called for. There is a deep freedom in that. [...] I don't need to believe that things will end well in order to act. The walking and the doing are their own reward. (222)

It's easy to agree with this; I guess many readers of Stoknes's book, as well as this one, will do so. That said, I shall end this discussion by returning to its starting-point: the disagreement we saw Stoknes begin with, between his mentor Jørgen Randers, a veteran of climate science, and himself.

Stoknes cites the bleak picture painted in Randers 2012 book *2052: A Global Forecast for the Next Forty Years*, that "Humanity will fail the challenge to act on climate change before it reaches runaway proportions, in the main because people, capitalism, and democracy are too short-term to tackle the critical long-term climate issue" (2015: xvii). Stoknes, half-seriously calling himself "the younger optimistic psychologist", is not convinced by Randers's diagnosis; he wants to "prove his dear colleague wrong" (xviii).

So how does Stoknes go about proving Randers wrong? He does so by way of accepting, as far as I can tell, the main premises of Randers approach, namely that what stands in the way of solving the problem is that "our thinking is too short-term and our behaviour too self-interested", prompting the question "Are we humans inescapably locked into short-termism?" (xvii).

Can we really be sure that these are the questions to ask? More to the point, that the assumptions they tacitly rest upon, unquestioned as such, are valid? Doesn't the sweeping claim that "people, capitalism, and democracy are too short-term" run together very different agents, levels of power, and dynamics of cause and effect? What exactly is short-termism, and what causes and sustains it? Is the short-termism said to characterise capitalism (these days, or always; inevitably or provisionally?) a product of the short-termism of the people making it up (say, the owners of the means of production, to allude to Marx); and if so, is that because of some deep-seated trait of human psychology, or is it rather to do with self-interest, or opportunism? And what about democracy in this equation? Does democracy appear committed to short-termism because of the individuals making it up, say, as voters; or does democracy in its current form, in allegiance with current capitalism, make up a culture that demands short-termism from "their people"? Is short-termism a bottom-up thing, or a top-down one? What is the role of culture? Which forms does power assume here, with special regard to addressing climate change? What kind of social forces – as distinct from individuals, very generally described – advance addressing it, and which advance blocking it?

In formulating this series of questions, I may come across as asking for more analytical clarity and for a more interdisciplinary approach than the one Stoknes sets out to offer, being a psychologist (who also happens to be an economist, though this background and expertise is only mentioned in passing, not as such drawn upon).

Nonetheless I do believe that the questions raised are worth asking to assess the validity of Stoknes's position. For all the complexity they introduce, the critical point I am making boils down to the following: Is the short-termism referred to be understood as a cause or as a symptom? Is it invoked as *what needs to be explained*, or as *what provides an explanation* (i.e., for the failure to tackle climate change over the last 40 years)? And as for the claim that "our behaviour is too self-interested" – is that what we have seen in the material cited above, in case stories like Norgaard's Eirik and Lertzman's Sally? I don't think so. Rather than their behaviour being too self-interested – or their being too self-interested explaining their behaviour – I think that both Eirik and Sally, to the extent that they exhibit denial, do so for very different reasons and motives. If anything, their stories are about attempts made to break out of the impasse and inaction, bordering on indifference that they perceive in their social environment. As we have seen, these attempts are met by hostility by their fellow residents, that is, by the silent or not so silent majority of people who hold on to their investment in denial even as – or precisely as – they are challenged to let go of it by individuals – such as Eirik and Sally – who are in the process of doing so. "Self-interest" is not what we encounter here; it is the wrong label to describe what is going on.

As mentioned, another major weakness I find in Stoknes is his failure to examine the dialectic between culture and individual, a failure revealing that a psychological inquiry of the very general kind he advances can only get us so far. True, at some points Stoknes does note that "cultural identity overrides the facts", taking this to shed light on "the greatest science communication failure in history: The more facts, the less concern" (81f.). Again, sweeping generalizations like this fail to provide a nuanced and complex way of understanding denial. Recall Norgaard's observation that "nations' per capita carbon dioxide emissions are negatively correlated to public concern", so that "the willingness of a nation to contribute to reduction in greenhouse gas emissions decreases with its share of these emissions" (2011: 77). Stoknes's failure to pay attention to this correlation – as distinct from the crude one between facts and degree of concern – is part of his failure to provide a systematic account of culture, that is to say, of *a particular culture* understood as what mediates, at the time and place in question, between facts and concern, or emissions and denial, informing the residents about how they are to interpret, communicate, and react to those facts, those emissions; they do not do so in a cultural vacuum, or in the same way in different cultures. The fact that Norway, a proud democracy, according to surveys has the same high degree of climate denial, roughly one-fourth of its population, as other major oil exporters and emitters like the United States, Russia and Saudi Arabia, speaks volumes about the impact of culture Norwegian-style in this era, an impact largely ignored, or should we say internalized and taken for granted, by the Norwegian psychologist.

We now turn to a systematic assessment of the role of culture thus understood.

Bibliography

Albrecht, Glenn, 2019. *Earth Emotions*. Ithaca: Cornell University Press.

Camassa, Manuel, 2024. *On the Power and Limits of Empathy*. New York: Palgrave Macmillan.

Cohen, Stanley, 2001. *States of Denial*. Cambridge: Polity Press.

Connor, Linda, 2016. *Climate Change and Anthropos*. New York: Routledge.

Cunsolo, Ashlee et al., eds., 2017. *Mourning Nature*. Montreal: McGill-Queen's University Press.

Freud, Sigmund, 1985 [1917]. "Mourning and Melancholia", in *On Metapsychology*, 251–268.

Goffman, Erving, 1959. *The Presentation of Self in Everyday Life*. Harmondsworth: Penguin.

Klein, Melanie, 1988. *Love, Guilt, and Reparation and Other Works 1921-1945*. London: Virago.

Klein, Melanie, 1989. *Envy and Gratitude and Other Works 1921-1945*. London: Virago.

Klein, Naomi, 2014. *This Changes Everything. Capitalism vs. the Climate*. London: Allen Lane.

Lertzman, Renee, 2015. *Environmental Melancholia: Psychoanalytic Dimensions of Engagement*. New York: Routledge.

Long, Susan, 2008. *The Perverse Organisation and its Deadly Sins*. London: Karnac.

Meadows, Donella, Jørgen Randers, and Dennis Meadows, 2005. *Limits to Growth. The 30-Year Update*. London: Earthscan.

Norgaard, Kari Marie, 2011. *Living in Denial. Climate Change, Emotions, and Everyday Life*. Cambridge, MA: MIT Press.

Orange, Donna M., 2017. *Climate Crisis, Psychoanalysis, and Radical Ethics*. New York: Routledge.

Oreskes, Naomi, and Erik Conway, 2010. Merchants of Doubt. *How a Handful Of Scientists Obscured the Truth on Issues from Tobacco Smoke to Global Warming*. New York: Bloomsbury Press.
Stoknes, Per Espen, 2015. *What We Think About When We Try Not to Think About Global Warming*. White River Junction: Chelsea Green Publishing.
Vetlesen, Arne Johan, 1994. *Perception, Empathy, and Judgment*. University Park, PA: Penn State Press.
Winnicott, Donald W., 1963. *The Maturational Processes and the Facilitating Environment*. London: Karnac.
Winnicott, Donald, 1980. *Playing and Reality*. Harmondsworth: Penguin.

3 Building resilience in a neoliberal society

3.1 Everyone is supposed to recycle

Mark Fisher's *Capitalist Realism* is one of those books that invite being read in one sitting. I may have been among its first readers in Norway in 2009, having ordered it based on a favourable review in *The Guardian*. Running to 81 pages, it's the fluency of the writing combined with the scope of its analysis that marks it, to this day, as not only a quick read but a most memorable one.

Taking as his point of departure the Marxist theorist Fredric Jameson's oft-quoted statement "It's easier to imagine the end of the world than the end of capitalism", Fisher, a relatively unknown British writer and teacher at the time, covers topics that leftist giants had spent decades grappling with, centring around the rise of neoliberalism in Western societies since the late 1970s. The ways in which consumerism has come to pervade virtually all areas of contemporary life, from professional to private, figure as a recurrent theme in sociologists like Jean Baudrillard, Richard Sennett, and Zygmunt Bauman. Yet Fisher, drawing on this job as a teacher – a precarious one, true to his subject – and remaining an outsider to the academic establishment until his tragic death (he committed suicide in 2017, at the age of forty-eight) was able to articulate the critique of twins neoliberalism and consumerism in concrete terms rich in examples, having a keen eye for how societal macro dynamics play out on the microlevel of everyday life.

The destruction of the environment is not among the issues for which Fisher's book is known, being mentioned only in passing and to illustrate how the role of the state has changed in the last forty years. Yet what little he has to say is spot-on.

Fisher draws on a paper by Campbell Jones entitled "The Subject Supposed to Recycle" to pose the question "*who* is the subject supposed to recycle?" Jones, Fisher observes, denaturalizes an imperative that it nowadays would be senseless, if not downright unethical, to resist: "*Everyone* is

DOI: 10.4324/9781003660743-4

supposed to recycle; *no-one*, whatever their political persuasion, ought to resist this injunction" (2009: 66). Fisher's point is worth quoting in full:

> The demand that we recycle is precisely posited as a pre- or post-ideological imperative; in other words, it is positioned in precisely the space where ideology always does its work. But the subject supposed to recycle, Jones argued, *presupposed the structure not supposed to recycle* [italics mine]: in making recycling the responsibility of "everyone"

structure contracts out its responsibility to consumers, by itself receding into invisibility. Now, when the appeal to individual responsibility has never been more clamorous – in her book *Frames of War*, Judith Butler uses the term "responsibility" to refer to this phenomenon – it is necessary to wager instead on structure at its most totalizing. Instead of saying that *everyone* – i.e., every *one* – is responsible for climate change, we all have to do our bit, it would be better to say that no-one is, and that's the very problem. The cause of eco-catastrophe is an impersonal structure which, even though it is capable of producing all manner of effects, is precisely not a subject capable of exercising responsibility. The required subject – a collective subject – does not exist, yet the crisis, like all the other crises we're now facing, demands that it be constructed." (Fisher 2009: 66)

The "responsibilization" Fisher refers to is a central topic in much contemporary sociology, inaugurated in Ulrich Beck's path-breaking book *Risk Society*, published in the wake of the accident at the nuclear plant in Tchernobyl in April 1986. What caught Beck's eye at the time, in the first wave of the neoliberalism promoted by Reagan and Thatcher, was a shift of enormous magnitude: the *shift of agency from collective to individual*, from societal structure to the employee, parent, or pupil who, precisely as a single individual, was made responsible for the tackling of structural problems now transformed into so many "individual challenges". To pick an example, unemployment ceased being described as a genuinely *political* issue, to do with the common good and a such to be handled by the government and the state, and instead was turned into something for which the affected individual him- or herself was responsible: make an effort to get a new job, re-educate or move if necessary, and treat others in the same situation as competitors for a good becoming scarce, and you will be duly rewarded for taking it upon yourself to get out of a situation it is up to you alone, not some overburdened nanny-state, to resolve. (The British filmmaker Ken Loach has caught this tendency in a number of his movies, most memorably in "I, Daniel Blake".) Thus the meaning of responsibility was redefined: it came to mean responsibility *to* yourself and *for* yourself (the Germans had to invent the words "Selbstverantwortung" and "Eigenschuld" to catch the shift); a demand internalized by me as what I need to attend to in my life, in my situation, cutting responsibility's ties and extension to others and *their* plight, such *other*-directedness having traditionally been its ethical essence, expected not only of individuals towards one another but in the priorities of the welfare state as well.

As Fisher is quick to add to the quote brought above, "the appeal to ethical immediacy…permanently defers the emergence of such a subject", that is to say, the collective subject he sees as called for given the ecological crisis. "The problem", he says, "is that the model of individual responsibility assumed by most versions of ethics has little purchase on the behaviour of Capital or corporations" (67). True enough, the appeal to individual responsibility in the now "old-fashioned" ethical sense of an unconditional demand to attend to the Other's plight articulated in a philosopher like Emmanuel Levinas, often invoked by the sociologist Zygmunt Bauman (1995, 2008), appears powerless in the case of structural crises, be they economic or environmental. Criticizing the tendency to individualize, privatize, and so depoliticize mental health problems, Fisher contends that "affective disorders are forms of captured discontent; this disaffection can and must be channeled outwards, directed towards its real cause, Capital" (2009: 80). Admittedly, this may come across as a veritable short-cut, reductionistic, and simplifying the causes of current psychological disorders. However that may be, the point Fisher wants to make is a different one: "The proliferation of certain kinds of mental illness in late capitalism", he continues, "makes the case for a new austerity, a case that is also made by the increasing urgency of dealing with environmental disaster" (80).

That the self-declared Marxist Fisher should speak out in favour of austerity may come as a surprise – austerity being the very policies for which the neoliberal regime is singled out for particularly harsh condemnation, policies with literally deadly consequences resulting from the shrinking of the welfare state and its provisions. But again, Fisher's point is more fundamental. "Nothing", he writes, "contradicts capitalism's constitutive imperative towards growth more than the concept of rationing goods and resources. Yet it is becoming uncomfortably clear that consumer self-regulation and the market will not by themselves avert environmental catastrophe" (80). The case he wants to make, then, is for a "new ascesis", being well aware that if, as many studies show, unlimited license leads to misery and disaffection, "then limitations placed on desire are likely to quicken, rather than deaden, it". Nevertheless, Fisher maintains, "rationing of some sort is inevitable". This being so, "the issue is whether it will be collectively managed, or whether it will be imposed by authoritarian means when it is already too late" (80).

Taking stock some fifteen years after Fisher wrote this, have we witnessed the "new ascesis" he proposed? Definitely not. No "new" austerity of the environmentally oriented one Fisher had in mind has emerged. Instead, there have been plenty of rounds of the old one following the downsizing of the "overburdened" welfare state, exacerbating the big overall shift that Fisher identified as a shift of agency – and thus also responsibility – from public to private, from structure to individual. The welfare state that has been dismantled since the 1970s is (was) a genuinely "social" state, Zygmunt Bauman points out, in that its aim, mandate and legitimacy, is to "promote the principle of *communally endorsed*, collective insurance against individual misfortune and its consequences"; it is "an institutional incarnation of the idea of community in its modern form – an abstract, imagined totality woven of reciprocal dependence, commitment, and solidarity", serving as an efficient and jointly cared for defence against the twin horrors of misery and indignity (Bauman 2008: 139, 141).

Even though Fisher would certainly share Bauman's worries about the adverse consequences of the dismantling of the welfare state, his main point is a different one, worth recalling: "Nothing contradicts capitalism's constitutive imperative towards growth more than the concept of rationing goods and resources" (2008: 80). It's not that Bauman disagrees with this notion; he would certainly welcome measures of rationing if need be. Yet his focus, I presume, would be on such rationing being *fair* to everybody involved, especially the weak and most vulnerable, people risking being condemned to redundancy and treated as "human waste" for failing to exhibit the productivity and can-do attitude that has become a condition of earning respect and recognition. The growth imperative in its environmental consequences is not Bauman's focus, and in this he is representative of the highly influential generation of sociologists he is part of (or was, he died in 2017), including Pierre Bourdieu, Richard Sennett, and Anthony Giddens.

But isn't the social state Bauman seeks to defend, in its hour of need, premised on growth? Surprisingly, the question is something of a blind spot in his many books devoted to critiquing neoliberalism in the era of "liquid modernity" and "liquid lives"; and again, in this Bauman is not alone. Indeed, for Marx-inspired socialists of various strands and across the generations, the ills for which capitalism has been criticized have essentially had to do with the lack of fairness in the distribution of power and of the material goods brought by an economic system driven by profit and growth. This critique has been preoccupied with retaining – continuing, even speeding up – growth while discarding the profit motive that historically has privileged the capitalist class and enabled its exploitation of the workers; whereby the exploitation of the natural world has not as such been singled out for criticism. The notion that environmental catastrophe is already happening and urgently needs to be averted to avoid the worst scenarios, alluded to by Fisher, is not stated as such in Bauman – nor the insight that this catastrophe is a direct consequence of the profit *and* growth imperatives of capitalism as we know it. To identify and seize upon growth as a *problem*, since at the root of the current crisis, rather than as what is needed to succeed in attaining just distribution of the goods produced and seen as tantamount to "material progress" as supposedly wished for by all parties, is the truly radical move that Fisher is making – at least when looked upon in the tradition of Western Marxism.

Shall we conclude from this that the ascetism and "new" austerity that Fisher talks about, wanting to promote it rather than critique it, is simply not borne out in the years that have passed since his book came out?

Such a conclusion would not only be premature; it would prevent us from recognizing an important trend that Fisher did not anticipate, at least not in its full implications. The boom of discourses on, and strategies for the implementation of, *resilience* is what I have in mind. What, then, is the connection between ascetism and austerity as advanced by Fisher and resilience as currently promoted? Can we say that the latter is the form that the former now assumes? But if there is a sort of connection – continuity – between the two, what about Fisher's explicit *political* agenda, his critique of capitalism neoliberal style, namely the crucial role of *growth* in creating a need for "rationing goods and resources", and so requiring a new ascetism? Furthermore, who would be the appropriate agents in this regard?

3.2 Resilience as ideology

There are no answers in Fisher's slim volume, only the couple of sentences cited above which mark the end of his analysis. However, in their study *Resilient Life: The Art of Living Dangerously* (2014), Brad Evans and Julian Reid explore the issues raised in a well-informed way, showing up a connection that neither Fisher nor Bauman helps highlight, namely that between neoliberal policies and "sustainable development". We shall see that the glue in this connection is resilience in its various forms.

What is meant by resilience? At its core, it refers to an agent's capacity to tackle adversity, to adapt and adjust to changes, especially unpredictable and erratic ones, in its environment(s), the natural as well as the social and cultural one. To "build" resilience is to take steps to be able to resist "storms" of various sorts that are bound to come, only it is hard to tell exactly when, how, and with what consequences. The need to build resilience is not only dictated by the onset, and most likely increase, of storms to face; it is also dictated by the conditions and traits that constitute vulnerability: the weaknesses of various sorts that render a given agent vulnerable in helping cause the degree of being exposed to the storms in question. Resilience, then, is about the effort made to minimize such exposedness, replacing weakness with strength in order to succeed in the most fundamental task there is: survival.

To put it like this may invite the impression that, being fundamental to human existence, resilience has always been part and parcel of what individuals, and the communities they make up, are concerned with and fully devoted to. In a sense, that is indeed the case: survival has always been the issue-cum-task *par excellence*, everything else being premised upon it. But this is a truism, and the point here is to recognize how resilience in the last few decades has come to play a different role than in previous times and cultures; a redefined and amplified role at that.

To appreciate what is involved here, consider this quote from Mark Duffield:

> A neoliberal biopolitics is more concerned to promote life that is resilient, able to exist on the edge of survivability, and adapted to uncertainty and surprise; a life that has abandoned trying to know the future and its associated prudentialism. Instead of providing a modernist *freedom from* the social pathologies of everyday life through, for example, normative regimes of inter-generational social insurance, resilience is more attuned to a neoliberal ethos allowing a *freedom to* embrace contingency as the essence of foresight and enterprise. [...] Resilience speaks to capitalism's constant transgression of limits, its inability to leave anything alone for long and the permanent revolution it effects in social and institutional life. Life is speciated according to the usefulness, irrelevance or threat it represents for the infrastructural and biospheric systems necessary for capitalism's widening cycles of reproduction, consumption and accumulation.
>
> (Duffield quoted in Evans and Reid 2014: 20)

In a short time we have been accustomed to speaking about resilience as an inevitable and universal "given" of existence; as something humans, individually and collectively, have always had to develop so as to survive. In the present neoliberal society, there a logical step from what we saw a sociologist like Bauman identify as the responsibilization of the individual, to understanding the task of building resilience as a core element of such responsibility, indeed as proof that the individual has got the message.

Yet as we saw in Duffield's statement above, there is more to the forms that resilience currently assumes than this. To Evans and Reid, resilience is "an ideological project" that unmistakably carries the signature of "biopolitics" neoliberal style. In making their claim, they remind us that the concept of resilience emanates from ecology. Here, resilience foregrounds the "buffer capacities" of living systems in general, pointing to their ability to "absorb perturbations" or "the magnitude of disturbance that can be absorbed before a living system changes its structure by changing the variables and processes that control behaviour". In this ecological sense, living systems develop "not on account of their ability to secure themselves prophylactically from threats, but *through their adaptation to them*" (2014: 30; italics mine). Thus they evolve *because of* systemic shocks rather than in spite of them, so that exposure to "shocks that register from the minor to the catastrophic" is not something that should be avoided as such. Instead, being exposed *and* changing – adapting – as a result of it is very much the stuff of which life – that is to say, survival – is made.

At this point, we start noticing the aforementioned connection between neoliberal policies and sustainable development. These policies are not restricted to the responsibilization of the individual referred to. Covering a much broader spectrum, indeed all sorts of living systems ecologically understood, resilience not only resides in, and is a task for, the individual organism but for entire societies, even the biosphere. As Evans and Reid observe, when the concept of "sustainable development" was coined in the United Nations' so-called Brundtland Report "Our Common Future" in 1987, it was intended to be applicable to the whole natural world, including every known species, as well as habitats of all kinds, thought to be vulnerable to the degradation wrought by economic development – with the "optimistic" twist that there is no necessary conflict between the imperative of economic growth immanent to capitalism (a label not used; to this day, and as a Norwegian, I have never heard Brundtland use the word "capitalism") and what is needed to secure "sustainable development"; indeed, growing the economy and safeguarding the environment and its "services" were said to be perfectly compatible, making for a win-win scenario (see Brundtland 1987; Vetlesen 2025).

What interests Evans and Reid is the recent uptake of the adaptive model of resource-management worked out by "resilience science" that all sorts of institutions are involved in. They mention the "securitization of the biosphere" expressed in the privatization and trading of the flow of "ecosystem services" maintained by intact ecosystems, redefining the depletion of ecosystems that is a feature of globalization neoliberal style as a global security problem calling for market-based solutions (putting a price tag on the services involved, etc.), thereby obfuscating and marginalizing the genuinely ecological roots of the problem. To thus "financialize" and marketize the problem of environmental destruction is to depoliticize it, pretending that the destruction is inevitable, a side-effect of goods produced and progress made, and as such cannot be prevented at its root, the growth-and-profit imperative of capitalism that is never called by its name.

This is my way of framing it, not that pursued by Evans and Reid. What I just called "depolitization", they regard as testimony to politization; and perhaps both characteristics are true to different aspects of the same process. Their main focus, then, is on two kinds of shift: in one move, resilience has shifted from "being a property of the biosphere to being a property of humanity", while in a second move, "service" has shifted from "being an element of economy to being a capacity of the biosphere" (2014: 35). As a result, development policies for increasing resilience, now considered a ubiquitous concern-cum-imperative, a must-do in so many areas, human and other-than-human, involving all sorts of entities and agents, have to be aimed at "ensuring access to markets". Originally conceived ecologically, as a property of ecosystems great and small on whose flourishing and intactness human societies depend, resilience has been extended to "a property which humanity intrinsically possesses, is capable of developing further, and which it can never have too much of". In this way, "a doctrine which started out as a critique of neoliberal policy prescriptions for development, ends up legitimating a neoliberal model of development based upon the constitution of markets and the interpellation of subjects within markets" (2014: 37).

If anyone believed that resilience would be about building *resistance* to current societal trends, or to ongoing ecological degradation, they have been proven wrong. Instead, resilience turns out to be a favourite tool to recruit ordinary people to play their loyal part in those trends, by adjusting and adapting to them as best as they can, all the while striving to become "the best version" of themselves qua individuals. Bauman speaks about "the insecurity of the present and the uncertainty about the future" that are "born of a sense of impotence; we seem no longer to be in control, whether singly, severally, or collectively" (2008: 26). "Realism" consists in giving up on any belief in the possibility to be secure on a permanent basis, holding security to be a once and for all attainable condition. Whether the world was ever such as to justify such an objective is not the issue. What matters is that security now becomes a life-long, never-pausing task, a permanent process of ongoing adaptation to dangers said and perceived to be outside the individual's control – and yet for the optimal tackling of which that very individual, vulnerable and exposed, is held responsible, whereby the proof of taking that responsibility seriously is by building resilience. To put it like this is to recall Ulrich Beck's original insight, mentioned above, about how individuals are instructed to take it upon themselves to come up with individual solutions to structurally generated problems.

The main argument advanced by Evans and Reid takes the form of a critique of naturalization: the rise of resilience as a strategy of adaptation, ultimately of survival, must be understood as geared to "naturalize the conditions of vulnerability" (2014: 65). "Sustainable development", they argue, "via its propagation of the concept of resilience, naturalizes neoliberal systems of governance and institutions". Thus, to insist on the necessity as well as the inevitability of building resilience is instrumental in advancing the entrepreneurial practices of self and subjectivity which are aptly articulated in the notion of "self-reliance". Of course, this fits hand in glove with what we saw Bauman describe as the dismantling of the welfare state as we had come to know it since World War II and up to the onset of neoliberalism in the 1970s and 1980s: people need to understand that their primary responsibility, especially in uncertain times like ours, is to take care of themselves in being prepared for the challenges ahead. Per necessity and "realistically", then, people need to abandon any old-fashioned idea about the government being there "for them", to protect them from adversity and to provide for their basic needs; to abandon, that is, the idea that institutions *per mandate* exist to compensate for the lack of self-sufficiency that affects all human individuals, since a universally given condition, non-optional and as such not to be eliminated.

To see what is at stake here, we need to determine what exactly is meant by "naturalizing the conditions of vulnerability". If I am not mistaken, what Evans and Reid are intent on criticizing is not "naturalizing" in the sense of neutralizing conditions of vulnerability, or casting them as inevitable, as a part of life being tough and so demanding people to be strong, not weak. Rather, the reason why naturalization is no innocent or neutral tool here is that the need to face up to the fact of vulnerability makes for something *positive*, not negative. This is brought out when the authors write that "exposure to disaster, in this context, is conceptualized in positive terms as constitutive of the possibility for the development of neoliberal systems of governance" (2014: 80). The possibilities created by having to come to terms with a world rife with dangers and disasters are eminently *translatable* into so many business opportunities as part of the market-friendly political regime of neoliberalism that Evans and Reid focus on. Thus understood, the situation is akin to the one thematized by authors such as Naomi Klein and Philip Mirowski in their books *The Shock Doctrine* and *Never Let a Serious Crisis Go to Waste*, respectively, going back to Milton Friedman's famous statement about "ideas waiting to be picked up" in times of crisis. The "shock treatments" said to be necessary typically take the form of austerity programmes tailormade to function – precisely – as so many opportunities for market-driven "solutions" provided by commercial and corporate players. This, of course, is a very different conception of austerity from the one propagated by Fisher above, meant in this case to enforce austerity, as part of a general "lowering of expectations" (Lertzman) onto ordinary people, especially those having the least and being hit the hardest, whilst the corporate players involved, for their part, aim for the opposite of austerity, namely expansion and growth.

What is being naturalized, then, is not so much the conditions of vulnerability as such, but the factors and forces that cause us, as individuals and societies, to be exposed to a world of dangers such as those brought by climate change. The causes that are at the root of the dangers are what are being naturalized – in a situation where, objectively speaking, these causes are anything but natural; they are eminently political, and they thrive, rather than being halted or slowed down, in the very regime that instructs the victims to be resilient and self-reliant. The root causes are the overexploitation and degradation of the natural world, as well as of human capital, that are the trademark of profit- and growth-driven capitalism. Perversely, instead of government and state protecting the people that are affected by the havoc wrought by these forces, it is the agents most invested in sustaining this self-same development that are being protected, shielded from the worst consequences and allowed to make money from continuing business as usual.

Evans and Reid bring an illuminating quote from an article by Henry Giroux, "Hurricane Sandy in the Age of disposability and Neoliberal Terror" written in the aftermath of Hurricane Sandy in 2012: "These days", Giroux writes, "Americans are quickly fatigued by natural catastrophe", adding:

> It gets worse. Catastrophes have not only been normalized, they have been reduced to the spectacle of titillating TV. Rather than analyzed within broader social categories such as power, politics, poverty, race, and class, the violence produced by natural disasters is now highly individualized, limited to human interest stories about loss and individual suffering.
>
> (Giroux quoted Evans and Reid 2014: 89)

In the wake of more recent hurricanes such as Helene and Milton in Florida in October 2024, what is striking is not only that Giroux's account still feels spot-on, but that the more frequent and in that sense statistically – and psychologically – "normal" these until recently extreme "weather events" become, the more important it apparently becomes for the media, including mainstream, *not* to speak about them as manifestations of climate change. Why? Only partly, it seems, because using that term, and so invoking the scientific explanation put forward by climate experts, would "frighten" people, or frighten them "unnecessarily", whatever that would mean. More to the point, climate change is not named as such because it is considered too contentious, too controversial; it would help reinforce a polarization within the public that has already gone too far. The more visible and in that sense undeniable the manifestations of global warming, the more the reflex of denial is stepped up. This does not only hold for the United States, in the grip of Trumpism; it holds for a far less polarized country like Norway as well.

According to Evans and Reid, the result of becoming resilient in the sense depicted

> involves the deliberate giving up of any possibility that climate change can be understood as a threat from which we can secure ourselves, as well as expectation that the state or any other political authority might be able to protect us from its dangers.
>
> (2014: 149)

This assessment fits well with what we saw Lertzman point out in the story of Sally above: how, that is, the individual concerned about climate change needs to learn the art of adjusting one's expectations and to live with disappointment. Whereas Lertzman's take on this focused on the individual,

Evans and Reid want to situate what an individual like Sally exemplifies within a broader political context, pointing to the "the system" that, as far as Sally is concerned, is an invisible entity – as if her world starts and ends with her home town Green Bay. It being invisible, or intractable, means that it is something that Sally may well sense that she is controlled by, yet as such is beyond her control. She has no way of bouncing back at the true drivers of the degradation affecting her and her place.

3.3 Climate science, a new religion?

The "futility of resistance", then, is common to the picture painted both in Lertzman and in Evans and Reid. But while Lertzman (2015) is content to leave it at that, by exploring Sally's reasons for giving up on resistance, Evans and Reid pursue the argument in a different direction: they want to critique what they call "the ecologization of the political on which neoliberalism thrives and the discourse of catastrophe has grown" (2014: 162). Climate science, they argue, is not a corrective to, or a way out, this predicament, but integral to it: "What does climate science express other than a longing for a sense of certainty; claims to truth that can be said to be beyond doubt?" There is a "scientific imaginary" at work that makes for "the impossibility of arguing with or over its [climate change's] reality", and this is an expression of such longing for certainty (162). Hence, claims to knowledge put forward by climate science as to "the coming of the 'Sixth Extinction'" go to show that we have to do with "a world populated by prophets that operate within regimes of truth deeply similar to those occupied by the prophets of Christianity. Climate science is a religion" (162).

To single out climate science for this kind of attack may come as a surprise in a book that clearly accepts man-made climate change to be real. Besides, to assert that climate science is a religion is to deliberately question the conventional view about the crucial difference between science as committed to truth (knowledge) and religion as an issue of faith. Why do Evans and Reid make this claim, and how do they justify it?

In a first move, they contend that

> climate scientists say that there is no way of escape from the dreadful and fearful realities of climate change; while economists say that there is no alternative to the further extension of the market in mitigation of the catastrophic effects of climate change.
>
> (163)

But how accurate and convincing is this parallel? To hold, as many climate scientists do, that the window is rapidly closing when it comes to implement the changes required to prevent the current crisis from worsening is meant as a plea for action, a call to arms (see Hansen 2009). To hold that "there is no alternative" to extending market-driven "solutions" is the opposite of such a call; it is Thatcher's TINA, meant to kill any ambition to change course, politically and structurally.

If this first move fails to convince, what about the second one? It consists in proposing that

> what is missing here is a different vocabulary through which to articulate the necessity and reality of climate change, while being able to welcome this inevitable event as the process of passage to a new world and new life beyond that which we have known until now.
>
> (163)

Indeed, they go on to ask, "who ultimately knows what the future for life is beyond the Holocene? Not one of us. The Anthropocene is only just beginning. What we can know is that life will take different forms" (163).

Why do Evans and Reid describe climate change as an "inevitable event", seemingly removing it from the sphere of politics, and so from the possibility of resistance that they in other parts of their book seek to promote? And why should this event be welcomed as "the process of passage to a new world", given the documentation they share about how disastrous this "passage" is, and will be, to millions of people around the world? It is difficult to view these statements as consistent with their overall argument about how resilience has become an ideology to oppose structural change of the kind necessitated by climate change.

In a third move, the argument becomes more specific. They point to the difference between "the life forms that will die off with the end of the Holocene and those that will emerge with whatever comes into existence after that time" 164). They invite us to consider, as an example, the "Grolar" bear, resulting from the cross between a Polar and a Grizzly bear made possible by climate change. Then they ask: "What is it about a civilization or a culture that manages to turn the wondrous phenomenon of the emergence of new forms of life, consequent upon these dramatic changes in a milieu, into a problematic of insecurity and threat?" (164). Quoting ecologist Brendan Kelly's observation that "this change is happening so rapidly that it doesn't bode well for adaptive responses", they see this as attesting to a "cult of mourning for the coming death of existing species life" and "fear for the nature of the new forms of life to come", a cult and a fear they consider to

> express perfectly the ways in which the ancient fear for the coming catastrophe is now coupled with a modern biopoliticized fear of the transformative effects of life's movement upon existing species. 'Pure' and 'native' forms thus become threatened by the emergence of impure, foreign, maladapted ones.
>
> (164)

These are remarkable statements. The named ecologist's warnings against the consequences for biodiversity on Earth likely to follow from unabated climate change and the destruction of habitats is being reduced to a prejudice in favour of "pure" and "native" forms of life; a quasi-psychological conservative defence in the face of the "impure" forms of life about to emerge. To embrace the "new" may appear subversive and in that sense attractive, sounding politically radical. But what about showing respect for the ecologist's expertise in sounding the alarm, based on knowledge and observations of current trends? 73 per cent of the world's wildlife – individuals, not species – have been lost in the last 50 years, according to the Living Planet Report published in October 2024; in my native Norway, 8 out of 10 seabirds have disappeared; the list goes on (see Vetlesen 2023: 202-251). The all-time-high decimation of biological diversity and richness is not about something "pure" being replaced by something "impure", as if the relationship between the two were proportional or symmetric. Instead, it is about an historically unprecedented impoverishment of life on earth, termed "biological annihilation" by a team of world-leading scholars (see Ceballos et al., 2017), exposing countless species to a vulnerability, a risk of extinction, for which the experts in the field say they are ill-equipped and ill-prepared. The introduction of alien species that, being impure and foreign (granted that these terms of theirs make sense), Evans and Reid are so fascinated by, poses a major threat to land-based as well as marine ecosystems around the world. Yet they are content to ignore this fact completely.

On Evans and Reid's view, to point out these facts based on ecological knowledge is to "submit to the blackmail of global ecological catastrophe", hence to "a combination of the very same modes of veridiction that functioned in the Middle Ages to subject human beings to absurd ideas such as the Kingdom of the Last Day and the Final Judgment" (165). If this is meant as an analogy between the present-day science of ecology and the way in which religion was used to "subject human beings to absurd ideas" in the Middle Ages, what precisely is it about the former that makes the analogy a valid one? What are the similarities? Evans and Reid offer no argument. It is as if they consider doing so superfluous, having stated that "climate science is a religion" (162). And by religion, we now realize, they imply not some sort of benevolent faith but the attempts of the powers that be in the Middle Ages, when the Church played a major role, to compel people to hold absurd beliefs that we nowadays regard as superstition, that is, as precisely the sort of ideas that modern secular science is meant to critique. Note that to assert that climate science "is" religion is different from saying that science is being presented as religion, or dressed up as religion. And indeed, some people sometimes do that (but Evans and Reid offer no names) – but not the scientists, and hardly a majority of those who follow the science and take it seriously.

There is something odd about this whole move on the part of Evans and Reid. To reconnect with my above discussion of denial: as we saw in the studies carried out by Norgaard (2011) and Lertzman (2015), many people – in countries like the United States and Norway, roughly one-fourth of the population – refuse to accept the findings of climate science. A considerable proportion of them regard the science as "fake", and castigate it is a "new religion", or worse, making their choice among the many conspiracy theories lying around like Milton Friedman's "ideas waiting to be picked up". If anything, the key among the obstacles that climate science faces is that many people dispute its findings, not that many – or too many – people look upon it as a new religion, and so as an enterprise beyond dispute and exempted from argument and documentation. Evans and Reid are clearly not talking about, or on behalf of, people like Eirik in Bygdaby or Sally in Green Bay when they ridicule those who have – we are given to understand – excessive confidence in climate science. What the case studies have shown is that taking the science seriously, and sharing the resulting discomfort with others, "kills the conversation"; it means social death, loneliness, the privatization of pain, rendering it invisible, depoliticizing what is a genuinely political issue concerning the common good. Again, the problem is not inflated trust in the science and in the reality of "catastrophe effects" (the scare quotes are Evans's and Reid's); on the contrary, it is the influence exerted by various structures and agents denying that science and its warnings, rendering denial – as practiced rather than intellectualized and actively argued for – the default position in many areas of everyday life.

If real-life individuals like Eirik and Sally are not the kind of people Evans and Reid have in mind, let alone the audience their book hopes to reach, who are? To answer this question, we need to go back to the authors' point of departure.

On my reading, the core issue here is to do with their view of *vulnerability*. Three distinct takes on vulnerability are discussed. The first is the definition alluded to above, employed in climate science, namely "the propensity or predisposition to be adversely affected" by climate change (used as a catch-all for environmental impacts), a propensity found in all sorts of entities, not only humans (2014: 4). Disaster management of the neoliberal kind, we are told, situates the vulnerable subject – taken for granted to be a human person (there is scant attention paid to other-than-humans in the book) – "within a political drama of catastrophe which renders it the author of its own planetary endangerment" (5). As we recall, this understanding of vulnerability in our days leads to a preoccupation, if not downright obsession, with strategies to reduce vulnerabilities, seized upon as so many business opportunities and pretexts to intensify the control and monitoring of the population. This, in turn, leads to what Evans and Reid speak of the effective blending of the terms resilience and resistance, a conflation that is ideologically potent in that it "signals the absence of any self-confidence in the liberal subject's disposition towards the world", a subject that "everywhere appears to be" – or is persuaded to perceive itself as – "under siege" by outside forces beyond its control. Yet their being outside its control does not mean that the individual subject is not expected to do its outmost to build resilience – quite the contrary.

The second take on vulnerability effectively locates what started out as a condition of all life in ecology and climate science within the current neoliberal society, turning a universal ecological given into a sociological and psychological one. This version of vulnerability is caught in Zygmunt Bauman's paraphrasing of Ulrich Beck's original insight whereby "individuals are expected to devise solutions to socially generated problems", so-called "individualization by decree" (2008: 47). The vulnerability of all individuals thus conceived necessitates the building of resilience within a neoliberal regime where the demand that individuals take responsibility in no way excludes "a massive role for state planning", so that the practices of resilience expand to straddle the private as well as the public, the personal as well as the political. Since risks of having one's "given" vulnerability exposed are found "everywhere", there is no limit to the agents and resources needed to handle it.

The third take on vulnerability is the most important one if we are to understand the position taken by Evans and Reid. They single out the prominent American philosopher Judith Butler as "the intellectual embodiment of vulnerability" (105). Human vulnerability – or precariousness, as she often says – is an ontological condition, one from which there is no escape and for which there can be no "solution". The attainment of security, understood as immunity to vulnerability-induced exposure, is but a "hyperbolic fantasy". In its stead, Butler promotes a new bodily ontology meant to provide the basis for a new form of leftist political subjectivity where the recognition that vulnerability is a "shared condition" will hopefully foster coalitions between different groups often perceived as political opponents.

Evans and Reid take issue with Butler's ontological understanding of vulnerability, both as a condition of human existence and of subjectivity and as a basis for a new leftist politics. Her way of thinking, they contend, is "contemporaneous not with an incipient new leftism, but with the dominant episteme and regime of power relations that the Left today has to combat" (106). Why? Because, the authors tell us, "the ontology of the social underlying Butler's account of vulnerability is deeply *liberal*" (107). To show this, they quote her statement that vulnerability "implies living socially, that is, the fact that one's life is always in some sense in the hands of the other", implying "a dependency on people we know, or barely know, or know not at all" (Butler 2009: 14). Their comment is that such statements go to show that Butler embraces "liberalism's understanding of the dangers posed by the subject's vulnerable life by the unknowability of the social". As opposed to this, what we should do is ask why, when, and how it was that the life of the subject came to be so conceived" (107).

This critique of Butler raises many issues and is itself problematic. First, the claim that Butler's way of understanding vulnerability is deeply liberal is outright false. The large majority of liberal thinkers from Locke over Kant to Rawls have been male, and the importance of becoming self-reliant has been a common theme, rendering vulnerability – to the extent that it is admitted as a "given" condition – a source of weakness, hence something to be hidden so as to prevent it from being exploited by others. In his book *The Self in Social Theory*, C. Fred Alford observes that for Hobbes "all activities of the self are oriented toward the reduction of vulnerability", and that for Rawls "justice becomes a matter of guaranteeing self-esteem against the onslaught of others" (1991: 120, 150). Whether or not Hobbes is to be regarded as belonging to, or being a precursor to, the liberal tradition is a matter of dispute (see Wolin 2004); not so with Rawls. The important point is that Butler's ontological understanding of vulnerability underscores an acknowledgement of *dependency* as fundamental to human existence, non-optional, and ineluctable at that, most dramatically so at birth, though never quite ceasing. The question of how to tackle this dependence on others clearly sets Butler's position apart from that of the liberal canon: whereas she asserts the need to embrace dependency-based vulnerability in human individuals as a core element – glue – of their relationships and bonds, liberal philosophers regard vulnerability as something to be overcome and minimized as far as possible, taking it as a source of weakness rather than as an invaluable and indispensable source of love, care, and solidarity, yet recognizing that these affirmative, other-directed emotions and ties are often fraught with conflict and ambivalence, humans being human (see Benjamin 1988).

To show how far away Butler is from the overly negative view of dependency and vulnerability that characterizes liberalism, consider this quote from the same page in her *Frames of War* that Evans and Reid are drawing upon:

> It is not that we are born and then later become precarious, but rather that precariousness is coextensive with birth itself (birth is, by definition, precarious), which means that it matters whether or not this infant being survives, and that its survival is dependent on what we might call a social network of hands. Precisely because a living being may die, it is necessary to care for that being so that it may live. Only under conditions in which the loss would matter does the value of the life appear. Thus, grievability is a presupposition for the life that matters. [...] Without grievability, there is no life, or, rather, there is something living that is other than life.
>
> (Butler 2009: 14, 15)

The second reason why their critique is mistaken is to do with the claim that Butler reiterates liberalism's view of "the dangers posed to the subject's vulnerable life by the unknowability of the social" (2014: 107). If, for the sake of argument, such unknowability is a trademark of liberalism, it is surely not so in the case of Butler. When Butler says that dependency on others, from birth and throughout life, is not restricted to "those we know" but includes many types of others, "most of whom remain anonymous" (2009: 14), she is not subscribing to a thesis about "the unknowability of the social" (whatever that may be); she is simply acknowledging a sociological fact, one whose truth Evans and Reid cannot – and try not – refute.

3.4 The politics of vulnerability

Why did I point out that the major thinkers behind the liberalism that Evans and Reid so mistakenly attribute to Butler's position are male? *Gender* is not a topic in their book; in Butler's – all of them – very much so. To state this is not only to point to a difference between her perspective and theirs that they should have paid attention to; more importantly, it is to raise the question why an inquiry into the various takes on resilience in contemporary neoliberal society can do perfectly without considering the importance of gender, of how notions of "masculinity" and "femininity" are drawn upon and play a part in how resilience is being experienced and assessed – not necessarily in the same way by girls and boys, women and men. Butler's stress on birth – and so on *giving* birth – is a case in point, ignored by Evans and Reid. Similarly, they overlook Butler's debt to Emmanuel Levinas (1985, 1991, 1993), the contemporary philosopher par excellence when it comes to how vulnerability issues an ethical demand "not to kill", that is, to accept unconditional responsibility to the other precisely because of the "weakness" perceived; this being the exact contrary position to that taken by Hobbes, for whom only strength – be it feigned, if need be – and the possibility of retaliation makes for the authority needed to protect oneself from one's vulnerability being exploited by others. Indeed, when Butler stresses "the fact that one's life is always in some sense in the hands of the other" (a statement stemming not from Levinas, but from the Danish philosopher and theologian Knud E. Løgstrup (1956), not mentioned by Butler), my impression is that to be reminded of this fact is not to Evans' and Reid's fancy. Why?

Let me put it like this: Butler is no stranger to Evans' and Reid's keen eye for the ways in which vulnerability, whenever exposed and wherever encountered, carries the risk of being exploited. We may go so far as to assert that the opportunities opened up for various agents – be they private, in the form of corporations; be they public, in the form of government and the state – by the exposure of vulnerability is a key feature of the neoliberal regime: namely, to seize upon the entities – the psyches, the bodies – thus exposed so as to profit from doing so, either in monetary terms or in terms of power, or both. That this is the case and has been so since the late 1970s is the main idea in Foucault's work on "biopolitics" and "governmentality", a main inspiration for Evans' and Reid's book, less so for Butler's.

Granted that they agree on this overall point about how vulnerability tends to be exploited in so many ways these days, what is it that they, more fundamentally, disagree about?

As indicated, I think their disagreement is to do with different views about, and perhaps experiences with, the condition of vulnerability. Gender is a part of this. Ever since Carol Gilligan's pathbreaking book *In a Different Voice* (1982), where, based on interviews with teenage boys and girls, she introduces a distinction between an "ethics of justice" and an "ethics of care", taking the former as articulated by the boys and as corresponding, roughly, to the position of Kant and Rawls (principles and rights are trump), and the latter as articulated by the girls and corresponding to the position (persons-in-relationships with particular bonds are trump) advanced by feminist scholars (tellingly lacking its own canon at the time), it has been clear that men and women are being socialized into holding different views and attitudes about the fact of vulnerability in human life, and in society's various institutions. I find this worth bearing in mind when interpreting what Evans and Reid say about vulnerability, either directly or indirectly when they critique Butler.

How to put this, more precisely? The two male authors are not interested in vulnerability *per se*; their topic is how it comes to be tantamount to weakness – potentially or actually – whereby such weakness needs to tackled, and the way to do so is by building resilience. And the more severe the societally produced environmental catastrophes of our time become, the more will the sites of vulnerability profligate, in turn amplifying the need to build resilience. The aim of resilience, its main task, is to neutralize vulnerability as far as possible. As Evans and Reid show in their book, the building of resilience takes on a novel shape in the current neoliberal regime, individualizing the task even while promoting a number of extra-individual agents to help attain the wished-for result: minimizing the costs of being vulnerable, neutralizing its potential for adverse consequences, stemming from it being a *risk factor*.

Whereas Evans and Reid and the authors they draw upon, such as Nietzsche, Deleuze, and Sloterdijk (needless to say, all male), take the understanding of vulnerability as a risk factor for granted, Butler for one does not. While not disputing that vulnerability is widely perceived as a liability, a source of danger and exploitation, in contemporary society, that this should be so is not part of any "original position" (a device employed by both Hobbes and Rawls, despite their – often overstated – differences); that is to say, there is nothing "natural" or inevitable to it. We saw how dismissive Evans and Reid are of Butler's Levinas-inspired ontological take on vulnerability as an ineluctable and as such universal condition of human life, immediately equating it "with the dominant episteme and regime of power relations that the Left today has to combat", that is to say, with an ontology that is "deeply *liberal*" (106f.).

For lack of a proper discussion, let alone argument, this is simply guilt by alleged association; and the association is misleading and mistaken. In fact, the Levinasian and feminist approach to vulnerability developed in Butler's books is deeply at odds with liberalism and the individualism it takes as "original" as well as normatively promotes. "The condition of primary vulnerability, of being given over to the touch of the other, even if there is no other there, and no support for our lives, signifies a primary helplessness and need, one to which any society must attend" (2004: 31f.). Far from resting content with highlighting this existential dimension of vulnerability, as such common to all humans, Butler goes on to critically address the *politics* of this dimension, namely how what is a shared condition of life is made into one shot through with inequality thanks to "the conditions under which certain human lives are more vulnerable than others, and thus certain human lives are more grievable than others" (30), creating a hierarchy of grief where some lives, or entire groups (women, blacks, Muslims, Jews, etc.), are not considered worth grieving at all, denying them in death the recognition denied them while alive.

By contrast, it never occurs to Evans and Reid to pay tribute to Simone de Beauvoir's famous critique in *The Second Sex* of how Western philosophy has portrayed the "human individual" on the tacit model of man, rendering the male way of being in the world the primary one and the norm, the female one secondary and derivative. Can it be that their account of vulnerability, and the way they immediately identify and associate it with weakness and exposure to exploitation, while aspiring to be universal inasmuch as no gender is given, or considered relevant to name, comes across as more true, plausible, and recognizable to men than to women? That women are more inclined to perceive and treat the vulnerability of the other – be it based on the experience of pregnancy and giving life – as the basis from which loving care and the desire to protect that vulnerable other spring? Vulnerability as a gift, something to act upon, someone to be there *for*, rather than representing a threat? What we encounter here is a desire to *offer protection* as a response to that person's *need to be protected*, as parties to an asymmetric relationship (Levinas), exemplarily parent and child? I am not suggesting that such a different approach to vulnerability and what it means – psychologically, morally, and politically – is a biological given; it is of course shaped by socialization, systematically so in ways that are liberating as well as suppressive and restrictive for both sexes, as Gilligan duly notes in her pioneering study. Whereas the culturally entrenched expectation that girls and women be more caring and doing most of the emotional work in family life may conflict with their aspiration for autonomy, the expectation that boys and men be independent and tough, too tough to cry, may conflict with their emotional needs and maturity.

Sociologist Thomas Scheff's seminal work on what he calls "the silence/violence pattern" is instructive here. He observes that in Western cultures "most boys learn, as first option, to hide their vulnerable feelings in emotionless talk, withdrawal, or silence". He terms these three responses "emotional silence": "Vulnerable feelings are first hidden from others, and after many repetitions, from self." As a result, men are more likely than women to show anger: "Men seem to be backed up on a wide variety of intense feelings, but sense that only anger is allowed them" (2006: 24). By way of example, Scheff recalls that implied in his father's comments and actions was a message about "equating fear with cowardice", instilling in the boy the need to gradually "numb out feelings of fear" to become the man that the father, representing society's values and norms, expected to see (25). Expressions of grief and fear, then, would indicate weakness, and so had to be suppressed to the point of not being felt in the first place. Such a highly selective pattern, dividing harshly between the few feelings (mainly anger) allowed and all those banned, comes with a heavy cost: "As we become more backed up with avoided emotions, we have the sense that experiencing them would be unbearably painful. In this way, avoidance leads to avoidance in an ever increasing, self-perpetuating loop" (26).

Theodor Adorno's reflections on hardness in his famous 1966 radio address "Education after Auschwitz" go to the heart of this topic. Clearly inspired by Freud's observation that our actions – especially those driven by strong emotions – have a two-dimensional structure in that when I do something to you, I at the same time do something to myself, Adorno describes the powerful idea that "whoever is hard with himself earns the right to be hard with others as well, and avenges himself for the pain whose manifestations he was not allowed to show and had to repress" (1998: 198). To counter this idea, Adorno maintains that what education must take seriously is that "anxiety must not be repressed": "When anxiety is not repressed, when one permits oneself to have, in fact, all the anxiety that this reality warrants, then precisely by doing that, much of the destructive effect of unconscious and displaced anxiety will probably disappear" (198).

To stress how the socialization of one sex differs systematically from that of the other may seem old-fashioned, capturing how those different expectations were internalized and acted upon in previous generations and not the present one(s). But this is not the case. Indeed, I think there are important differences in how men and women in present-day society relate to climate change – by which I mean not only how they relate in terms of knowledge, but how they relate emotionally as well, with girls and women often being criticized – mostly by men – for being "too emotional", as opposed to being "rational", the two being considered as mutually exclusive (I shall say more about this in Chapter 4 when discussing Greta Thunberg). The mantra of economic growth, endless growth, is in denial about human societies' dependence on nature, about the limits this implies, not only ecologically but psychologically as well.

Reminders of dependence, be it on other persons, be it on the natural world, conflict with the trump (or nowadays: Trump) "masculine" values and ideals of neoliberal society, namely freedom and independence, where freedom is regarded as premised upon gaining independence and sustaining oneself – surviving – as self-reliant and self-sufficient as far as possible. Thus understood, freedom means "overcoming all limits to my will", whereby freedom is defined as "possession of money and power: money buys freedom; freedom means having enough money to do what I want", to cite C. Fred Alford's study on how freedom is conceived by a majority of Americans (2005: 1, 49). Traditionally, men have been given more opportunity to pursue this ideal than women, involving a trajectory from initial dependence on others to increasing independence from them, understood as the attainment of freedom (doing what I want without being restricted by others, or by big government). Whether this ideal should be made truly universal, meaning genuinely applying to both sexes, or whether it would be more progressive to challenge the hegemonic view about dependence and freedom as "inevitably" contradictory and at odds with each other, or rather complementary and mutually supportive, is a big and ongoing debate.

My claim here, relevant to exploring the phenomenon of climate denial, is that a macho worldview is bound to be deeply offended by reminders of dependence and its link with vulnerability – in oneself, and in the society one belongs to. That mother nature has been exploited for so long, and so mercilessly, in the name of freedom and prosperity, only now to bounce back with a vengeance – payback time – is so out of tune with this world-view, and something one is so culturally and psychologically unprepared for, that strategies to deny it are eagerly adopted, in a situation where such strategies are in no short supply; indeed, for parties offering them they prove a recipe of electoral success, again and again, especially with male voters, the young as well as the old.

Donald Trump is a case in point to show how the macho worldview I am describing plays out, manifest in his denial about the climate crisis and his demonstrative indifference to environmental destruction and its victims. As described in the book *Too Much and Never Enough*, by Mary L. Trump, a clinical psychologist and Donald's only niece, from an early age Donald was taught by his powerful and authoritarian father Fred that "weakness was perhaps the greatest sin of all", echoing *verbatim* the observations made by Scheff and Adorno, quoted above, thus serving as a paradigm example of the very sort of socialization that Adorno warned against. Fred wanted his son to be a "killer", and "being a killer was really a code for being invulnerable" (2020: 41). Donald witnessed how his older brother Freddie was "too soft" to gain the respect of the father, taking it upon himself to prove his worthiness as the one "true" son, to be trusted with taking over his father's business empire. The father's "fundamental beliefs about how the world worked – in life, there can only be only one winner and everybody else is a loser (an idea that essentially precluded the ability to share) and kindness is weakness – were clear" (43). To protect himself and appear always strong, never insecure, Donald learned how to keep most people at arm's length. "Life in the [family] House", Mary Trump writes, "made all the children in one way or another uncomfortable with emotions – either expressing them or being confronted with them" (44). The experience of leaving home and being enrolled at NYMA, a private boarding school, "reinforced Donald's aversion to vulnerability", which, she notes, "is essential for tapping into love and creativity because it can also expose us to shame, something he could not tolerate" (53).

Mary Trump's portrait of her uncle can be enriched by invoking what Melanie Klein (Klein 1989) refers to as the "paranoid-schizoid position". In this position, based in early childhood and as an attempt to tackle the radical dependency on another (usually mother), we fear for ourselves and all that is valuable to us at the hands of what is perceived as malevolent external persecutors intent on destroying us. Klein stresses that these persecutors are of our own making, only we are unprepared to admit it; they are our own projections, our rage, jealousy, and envy come back to haunt us, spelling, C. Fred Alford (1997: 40) remarks, the psycho-logic of *lex talionis*, in which we "fear everything we would have done to others". Note how readily evildoing may be enacted out of paranoid-schizoid anxiety: "If one experiences the world as the source of perpetual attack, the best defense may be to attack first" (1997: 40). As mentioned earlier, Alford considers Thomas Hobbes the philosopher who best describes this logic, showing (in *Leviathan*) how pre-emptive attack is the best defence in the state of nature – the state a person like Trump ever since his childhood, and later due to the influence of Roy Cohn, takes to be "just under the surface" of present-day society, its "truth" and so what one can ignore or wish away only at great peril.

In this scenario, which continues from childhood and may last throughout life, and so is never abandoned, modified or followed by greater maturity, the fundamental fear is not fear of death but fear of narcissistic injury, nourished as it is by an extreme vulnerability to invasion and appropriation; a vulnerability, moreover, not easily overcome because there is no clear (robust) distinction between self and world; it is as though "the whole world were but a mirror of one's own passions", the world in general, and the people with whom I interact in particular, being perceived only in terms of how it makes me feel (see Alford 1991: 94ff.; Vetlesen 2005: 118ff.). Accordingly, in the paranoid-schizoid position, "every act of care and concern is purchased by an act of (or phantasy, which rebounds as fear of persecution: talion morality) of aggression and hatred against others", the logic being that "the more we love someone, the more we must hate others, in order to protect the loved ones from our own aggression" (Alford 1989: 88f.).

In my view, many of the features of Trump's personality and his style as a political leader can be illuminated by referring to Klein's paranoid-schizoid position – features such as splitting, perceiving things and persons in a black-and-white manner, in terms of either/or, disallowing for complexity and ambivalence; absence of feeling shame, of being ashamed of oneself and so of feelings of guilt, regret, and wish for reparation (the latter characterizing Klein's "depressive position"); the denigration of empathy, seeing it as tantamount to weakness and so as despicable, "unmanly"; the obsession with revenge, enacted as vindictiveness against enemies, actual or fictitious; the manifestation of envy, never admitted as such because it would betray inferiority; envy understood as the attempt to *destroy the good itself*, as distinct from greed, which aims for possession of all the goodness – value – of the object, spurred as envy often is by sheer spite: if I cannot have all the good myself, then no one else shall have it either, especially not people considered to be "good". If the good is successfully destroyed, then there is no reason to feel the discomfort of envy (see Alford 1989: 37). What is truly radical about envy is that it ultimately seeks to empty the world of goodness. In Klein's view, this makes envy "the worst sin there is; for all other sins are sins only against one virtue, whereas envy is against all virtue and all goodness" (Klein 1989: 189; see Alford 1988: 33). I shall return to the issue of wishing to destroy goodness, and the seemingly opposite phenomenon, that of denying one's own as well as other's capacity for goodness, when discussing leadership in Chapter 4.

To write about Trump in this psychoanalytic manner may seem an aside here. It is not. The toxic family life he grew up in helped make him the man he was to become, an embodiment of toxic masculinity, shamelessly, and defiantly so. This would not be relevant for my discussion were it not for the fact that for so many men, Trump is a hero, a courageous leader onto whom self-image (Freud: ego ideal) and admiration are projected. His ascendency to the heights of power – President of the United States, elected twice – is literally borne by the widespread perception – among men – that he symbolizes "payback time" on a whole series of issues, ranging from political correctness (WOKE), to feminism and environmentalism. The sheer claim that as humans and as a society we are dependent on nature and its flourishing, that we are at the receiving end of this relationship which is asymmetric and with us (meaning men as well as women) being the weakest part, at the mercy of sources of goodness and gifts we cannot control, is intolerable to the macho world-view Trump epitomizes. To put it like this is to point to an undercurrent of Trumpian climate denial that goes much deeper than "not believing the facts" presented by the science, or being in inner conflict because of inconsistency between attitudes and actions, what is said and what is done (to allude to my discussion of Per Espen Stoknes (2015) in Chapter 2). To the contrary, it is the presence of *consistency*, defiantly held on to and stubbornly acted upon, not its lack that is conspicuous here. Driving big fossil cars and chanting "drill, baby, drill" are not practices creating conflict with the underlying worldview and its trump values of individual freedom and independence, but a matter of playing them out for all to see, with nothing to hide or be ashamed of. As we will see below, for all its consistency, this type of masculine mindset comes with a considerable cost, personally as well as politically.

3.5 The Nietzschean inspiration

After this discussion of the significance of gender, let me return to my discussion of Evans and Reid.

I mentioned that they draw on Nietzsche, and the way they invoke his work is instructive for assessing their understanding of the pair vulnerability-dependence, and its counterpart, as it were, in neoliberal society, freedom-independence, as set out above. This society encourages its members to view the former pair as overly negative, the latter as positive, making for a conflict between them that requires being handled on both an individual and a societal level. When we ask why there is, or needs to be, a conflict, the answer is that there is an either/or, not a both-and, such that vulnerability-dependence represent an obstacle to freedom-independence, zero-sum like: the cherished values of the latter can only be attained and secured by neutralizing the impact of the former.

How does Nietzsche enter this picture? Evans and Reid are interested in his account of nihilism, whose power in the modern world they see as deeply characteristic of it. Nihilism refers to the triumph of what Nietzsche called "reactive thinking", namely a radical repudiation of value, meaning, and desirability. Spontaneous affirmation of life was thwarted, as was any manifestation of creativity and aggressiveness, of being "hungry for life" and the enjoyment it may bring, sensuously, corporally, and intellectually, as so many dimensions of vitality and outward-seeking energy. In the guise of the "slave mentality" described in his *On the Genealogy of Morality*, all such manifestations of vitality are condemned, moralistically dismissed as egoistic and dangerous, hence at odds with the virtues of Christianity and any ethics worthy of the name, pity in particular. To the extent that this slave morality wins out over its rival master morality, the former suited for the many, the latter for the few, the result is nihilism, meaning the triumph of a will to nothingness that strips life of everything that makes life worth living in Nietzsche's affirmative sense. Nihilism as acted-upon leads to a feeling of impotence, the price to be paid for the internalized suppression of spontaneity and creativity, an impotence that spurs attacks on those (few) who still exhibit the life-affirming traits and aspirations now prohibited. This is the basis for *ressentiment*, blending fear, envy, and vengeance. Evans and Reid point out that in our day, the German philosopher Peter Sloterdijk equates this ressentiment with "rage, the basis of all great theisms" (2014: 116). In his late writings, for example, in the figure of Zarathustra, Nietzsche seeks to counter the cultural triumph of slave morality where humankind only diminishes itself, partly abandoning, partly denying its true potential for creativity, positing in its place a vision about life as a work of art, giving pride of place to a playful, sensuous, and in a broad sense aesthetical way of being in the world. Mentioning Gilles Deleuze in addition to Sloterdijk as a supporter of this vision, Evans and Reid refer to the Nietzschean idea of a poetic subject such that "he himself is really the poet who keeps creating this life" (Nietzsche quoted in Evans and Reid 2014: 172).

Why is this part of Nietzsche's thinking important to Evans and Reid, exploring as they do the role of resilience in contemporary society? Nihilism, they contend, "has never been alien to liberal biopolitics. It is arguably its most potent expression" (2014: 117). In keeping with their overall argument, the liberal subject is the resilient subject; a subject, they suggest, that no longer "solely projects its resentfulness onto the souls of "Others"; it "resents the living world, for it too is radically endangering". This subject is shaped and mobilized by the prospect of the coming catastrophe; it

> fears the transformation of the subject, just as it fears the transformation of the ecosystem that gives sustenance to life. Our rage as such, to borrow from Sloterdijk, has become truly limitless. As everything becomes the source of our endangerment, we internalize the *ressentiment* and proliferate our impotence with unrivalled intensity and absolute necessity. Hence this produces a form of nihilism which is 'unbounded'.
>
> (118)

I do not think this is a compelling argument. Evans and Reid rely upon the rhetorical force of suggesting links, making for affinities, for example, that between nihilism á la Nietzsche and liberal biopolitics. But to say this is not to propose that Nietzsche is irrelevant to the topic of resilience. Rather, to question the authors' way of invoking Nietzsche is to ask what kind of ammunition (to use a very Nietzschean word) one can get from turning to his account of slave morality and how it leads to nihilism.

In his book *Twilight of the Idols* (1970 [1889]), Nietzsche proclaims: "What does not kill you, makes you stronger". He continues:

> For what is freedom? That one has the will to assume responsibility for oneself. That one maintains the distance which separates us. That one becomes more indifferent to difficulties, hardships, privation, even to life itself. That one is prepared to sacrifice human beings for one's cause, not excluding oneself. Freedom means that the manly instincts which delight in war and victory dominate over other instincts. For example, over those of "pleasure" The human being who has *become free* [...] spits on the contemptible type of well-being dreamed of by shopkeepers, Christians, cows, females, Englishman, and other democrats. [...] how is freedom measured in individuals and peoples? According to the resistance which must be overcome, according to the exertion required to remain on top. [...] first principle: one must need to be strong – otherwise one will never become strong.
>
> (1970: 542)

There is little here that, say, Donald Trump would disagree with. This is the version of Nietzsche that may well have been on Ayn Rand's mind, by way of inspiration, when she wrote *Atlas Shrugged* and *The Fountainhead*, the two novels of hers that Trump boasts of having read as a young aspiring businessman; indeed, the ideological Bible of anti-government libertarianism. More to the point, there is also little here that would not qualify, splendidly so, for being cited in self-help books about how to build resilience in a dog-eat-dog society like our present one.

What follows from this? Is Nietzsche on the wrong side of the argument that Evans and Reid want to make, in their avowed critique of resilience neoliberal style? Even though the exercise of applying Nietzsche's formulations to current phenomena like resilience may, admittedly, be somewhat strained, this is what Evans and Reid, like Sloterdijk before them, appear happy to do.

Without mentioning him, I think the last paragraphs of their book owe a lot to the inspiration they take from Nietzsche, stating that their aim has been to "show how discourses of vulnerability, resilience and adaptivity cannot provide for a form of political subjectivity capable of outliving, meaningfully, the liberal subject" (2014: 202). Yet, as we have seen, to imply as they do here and throughout their book that neoliberalism actively inflates and exaggerates the human condition of vulnerability – only to exploit it – is, at best, to present a very selective picture of the state we're in. Yes, the commercial and corporate agents that have an interest in profiting from the levels of anxiety generated by climate change and precarity in times of austerity are real and thriving. But they are not engaged in *idealizing* the pair's vulnerability-dependence, or, as part of that, in portraying it as a state from which there can never be a real escape, or a true and lasting relief. Indeed, the whole point of building resilience – its *raison d'etre* – is that it seems within reach for the "responsible" individual to fight oneself free from the ways in which that pair come to hinder and slow down the attainment of the other pair, the truly desirable and valued one, that of freedom-independence.

By contrast, when Nietzsche singled out slave morality for his damning critique, he did so because he took it to idealize vulnerability-dependence, with its "female" affinities (mentioned in the above quote), and to deny the truth – increasingly precarious and to be recognized only by the few – about it, namely that it is possible to leave it behind, to transcend it in the pursuit, and possible realization, of "life as a work of art". If vulnerability-dependence is held to be a matter of *fate* according to slave morality, and so nothing worth striving to break free from, there being nothing "higher" to attain, neutralizing its negative impact on the individual's life as currently lived in these uncertain times is what the ideology characteristic of the neoliberal era is up to – and the contrast between the two takes, and assessments, of vulnerability-dependence is undeniable, much more important and consequential than their (alleged) similarities.

Given this contrast, what are we to make of Evans' and Reid's concluding statements – that "the liberal image of the human, degraded and incapable of action and meaningful creation, is the real chimera"; that "liberal regimes are putting so much effort into imagining the necessity and possibility of the resilient subject, equipped only ever to adapt to a world outside its control" (2014: 202f.)? Now, to the extent that resilience neoliberal style functions as an ideology, it does so by persuading the individual that if it obeys the demand to assume responsibility for its own success and survival, as it must and should, *that effort will make a difference*. Of course, Evans' and Reid's ideology-critical response to this is to say that this promise, this aspiration attributed to the individual, is false; it creates the impression that the individual can do more to combat the real reasons for the world becoming increasingly dangerous than what is actually the case.

Is it true that liberalism presents the human as "degraded and incapable of action"? There is not one, definite answer to this; it depends on whom among the (predominantly male) liberal philosophers you consult, as C. Fred Alford shows in his book (1991). Yet we should not complicate the issue unnecessarily. The "liberalism" Evans and Reid refer to is the real-life neoliberalism whose contemporaries we are. So how does the call for resilience fit in? In view of what we have discussed, we can say that the demand to become resilient makes sense because as individuals we are vulnerable and dependent, hence at risk of being exposed to dangers that may prove lethal, at worst. And yet, at the same time, to require of individuals that they make the effort to build such resilience only makes sense if they believe that the risk factors can be *thereby* handled, meaning minimized or effectively overwon. Note that the contemporary notion – and practice – of resilience combines both aspects: we have not chosen to be vulnerable and dependent beings (Levinas, Løgstrup, Butler), nevertheless, there is something we can do about it. According to one view, that of the thinkers just mentioned, the realistic and morally prudent way to relate to this condition of ours is to safeguard and protect it as something essential to who we are, individually and in our relationships. This is the basis for Schopenhauer's (1986) ethic of compassion ("Mitleidsethik"), comparable to present-day feminist ethics of care (see Vetlesen 1994: 206ff.). According to the other view, advanced by neoliberalism, vulnerability-dependence is a nuisance through and through, a source of exposure that is equal to weakness, a weakness that, if revealed, will immediately be taken advantage of by others; and the more so, the more precarious social life becomes, be it as a consequence of austerity brought by economic crisis, be it as a consequence of the dangers caused by climate change.

3.6 Two versions of vulnerability

What, more precisely, is the relationship between vulnerability and politics? What is the proper framework for exploring this? Is it a matter of the relationship between micro and macro, that is, between what is played out in and among individuals, and what plays out in institutions?

Of course, this is a much too crude way to make the distinction, but it may serve for analytic purposes.

Consider, for a start, the following features of vulnerability. Whereas all human individuals are vulnerable, some are more vulnerable than others. And whereas all qua vulnerable beings need to build resilience, some will be more successful in doing so than others, for reasons to do with resources – within and outside them – that greatly vary. That the resources necessary for building resilience vary among individuals, in systematic ways and patterns at that, tied to class, is what makes the constellation vulnerability-resilience *political*, a matter of power and its distribution. To say that it is political is, of course, to say the obvious. Yet, since the currently intensified demand to build resilience does not originate in the individual, but is addressed to it from the society it belongs to, culturally mediated at that, the resulting individualization of the demand conceals its political origin – critics would say, deliberately so and by design. Your failure to be resilient is yours and yours alone.

Again, this is not to say something that, say, Butler does not acknowledge. Even so, questions about the adequacy of her approach must be asked, questions to do with its political implications in particular.

I think that Butler, following Levinas, is right to conceive vulnerability-dependence as an ontological feature, as such common to all human beings regardless of their particular circumstances and environments. Now, whereas vulnerability as something shared underpins moral values such as empathy, equality and solidarity, resilience as a strategy for coping with vulnerability so as to thrive and survive, to not lose out because of it, entails competition: it is not only that resilience is itself a scarce resource; it is also a coping strategy in a world – social as well as natural environment – fraught with scarcity and adversity. This means that vulnerability and resilience point in opposite directions, as it were: the one in that of an ethic where responsibility is not simply, or even primarily, for oneself; it is for the other (Levinas, Løgstrup), asymmetrically at that, unconditional in the sense that I am addressed by your vulnerability, and the more vulnerable (and dependent) you are – "child, widow, stranger" (Levinas) – the more grave my responsibility. Alternatively, the ethic called for is an Hegel-inspired one of recognition, highlighting the possibility if not actuality of reciprocity over asymmetry (Hegel 1991; Williams 1997).

It is not clear to me which of these two versions of such an ethic Butler favours; sometimes the one, at other times the other. That is not the main issue here – instead, it concerns the implications of such an ethic for politics.

A book by the German sociologist Stefanie Graefe is helpful here. In *Resilienz im Krisenkapitalismus*, Graefe introduces a distinction between vulnerability understood as ontological, and as such as elementary and irrevocable in human life, calling it "Verletzlichkeit" (hurtability), and vulnerability understood as "Verletzung" (harm) as something concrete, something a particular individual is affected by in particular situations or circumstances. Graefe finds that Butler fails to offer a criterion to distinguish between being a vulner-able person (something to be affirmed and safeguarded) and being a person who is – now, here – being harmed (something to be condemned, and put an end to) (2019: 188). Whereas the former case is about being entitled to protection in virtue of being, ontologically and irrevocably, vulner-able, the latter is about instances where that condition is not respected but violated, something that as such is avoidable, should not happen, and makes the person violated in his or her vulner-ability into a victim. Put in my terms: whereas the former case serves as the foundation for ethics (be it Levinas or Løgstrup, be it Hegel), the latter points to politics, concerned as it is with the question of what to do, as a society, to see to it that violations of the said kind not occur, and be sanctioned if they do (the mandate of law).

Graefe's claim, then, is that Butler fails to make this distinction, and that her argument for the political implications of her view of vulnerability suffers for it. Is there not, Graefe asks, a danger that vulnerability gets ontologized, and that this leads to an ontologization of power and domination (2019: 193)? One could add: how to distinguish between vulnerability as *conditio humana*, unavoidable qua condition of existence, and the degree to which concrete persons are vulnerable, some much more than others, for reasons and in ways that are avoidable, rendering such instances criticizable in political terms, and sanctionable in legal ones?

I shall not try to answer what the sort of criterion Graefe is after might look like in Butler, given that – and I agree – she does not systematically provide it in her books. Graefe's more specific criticism is worth examining, though, namely that Butler "fails to show how vulnerability can be overwon, and how security is possible and can be provided, be it thanks to social institutions" (2019: 192). She finds that the institution – or institutionalization – of human rights is looked upon with suspicion by Butler, even though her propagation of a principle of equality, based on the vulnerability common to all human beings, crucially relies on institutions in order to be upheld in practice in a modern and complex society like ours. Indeed, Graefe goes on, for lack of appropriate institutions, the scandalization of instances of violation of concrete persons' vulnerability, the outcry, rage, and protests they may lead to in the particular case (the death of George Floyd springs to mind), would come to naught and be testimony to impotence *politically* speaking.

I agree that Butler fails to offer the kind of account of institutions that Graefe is after in Butler's books *Precarious Life* and *Frames of War*, although it must be said that the latter does offer an extensive critique of the role of the military in the human rights violations carried out in Abu Ghraib and Guantanamo, and of immigration authorities under recent American presidents of both parties. What Graefe *is* right about though, is the more general, and I would say deep-seated, suspicion towards institutions – economic, political, legal – that is evident in Butler's work. This suspicion is borne out by the fact that when she does examine the workings of institutions, the cases, and the examples selected are negative, going to show how the tasks and powers entrusted to them – be it upholding human rights – are abused, time and time again, sustaining a politics of inequality and injustice.

We have been here before, albeit from the opposite vantage point, as it were. Discussing Gardiner's work in Chapter 1, we saw how he set out to highlight the role of moral corruption in the failure to tackle climate change. However, as is typical of scholars coming to this topic from a liberal point of view, Gardiner's take on corruption proved deficient and selective: he concentrated on how corruption plays out in the individual agent, and failed to examine its impact *within institutions*, and so to raise the disquieting question whether corporations – not all, but some of them – are pervaded by corruption, deceit, and denial, such as to have become a systemic feature rather than a sort of moral failing, or vice, found in some individuals but not in others.

If Gardiner's trust in institutions seems naïve, simply taken for granted as liberals committed to the present capitalist society are wont to do, the situation with Butler is the reverse: a proponent of the radical left, she is committed to the sort of "hermeneutics of suspicion" that is associated with Marx and Freud, showing up so many social practices and political-economic institutions as carriers of systemic inequality and injustice, whereby "Western values" such as human rights, dignity, and freedom for all, regardless of group affiliations and identities, are to be looked upon with deep distrust, through the critical lenses of *Ideologiekritik* of the sort pursued by (early) Frankfurt School critical theory. Corruption among individual agents there may well be, exhibited in too many cases to mention, with fatal outcomes for the victims, yet all-too-often with impunity for the offenders. But the problem is systemic not individual. The policeman Derek Chauvin's murder of George Floyd in May 2020 (mentioned above) is a case in point, where "the larger picture", Butler writes in *The Force of Non-Violence*, "is one in which states and legal powers justify their own violence as legitimate coercion and cast all forms of counter-violence as unacceptable violence" (2020: 123). I shall come back to the latter aspect below, when discussing how protests against climate change are being met by relevant authorities and institutions.

Butler observes that

> the condition of interdependency that makes our lives possible can also be one condition for exploitation and violence. The political organization of life itself requires that interdependency – and the inequality it implies – is acknowledged through policy, institution, civil society, and government.

She goes on to say that

> to oppose violence is to understand that violence does not always take the form of the blow; the institutional forms through which it operates compel us to ask: Whose life appears a life, and whose loss would register as a loss?.
>
> (46f., 63)

The question is how deep Butler's suspicion of current institutions runs. Of course, there can be no general answer to this; it depends on what kind of cases we consider. There is, for example, Chauvin's murder of Floyd, a case that spurred Black Lives Matter to demand the "defunding of the police", a demand that, at the time, achieved a majority vote in the city of Minneapolis, where the murder took place. The radical left stance was that Chauvin's misconduct was not one bad apple but an instance of a piece of behaviour seen a thousand times before, manifesting a pattern, *systemic to the institution* – the police – itself, sustained for decades by way of being culturally entrenched among generations of policemen, many of whom are themselves black. It doesn't help that the offenders (sometimes) have the same colour as the statistically over-proportionate victims – not if the system is broken.

Butler is aware that discourses of "vulnerable groups" risk reproducing paternalistic power and giving authority to "regulatory agencies with interests and constraints of their own" (2020: 185). And she acknowledges that we are not in a position to "establish vulnerability as the foundation for a new politics" (186). The limitations of an ethic of vulnerability considered as a basis for politics have been pointed out many times before, going back to Levinas's difficulties with transposing an ethic of the "moral party of two" (Bauman), where the agent's dedication to the Other qua face is unconditional, to a politics where "the Third", and sometimes countless, faceless, non-encountered others demand justice. To enter the terrain of politics means to leave the singularity of the individual behind and confront a plurality of individuals, hence to engage in comparison between the many "needy" in situations where resources are scarce, making for an "economy of quantity" far removed from the unquestioned primacy of the Other from which ethics starts (Levinas 1993; Vetlesen and Jodalen 1997).

Butler does not discuss the transition from ethics to politics in detail. She is much more interested in the role played by the constellation power-violence in the concrete case, being content to state that "considered as a condition, vulnerability can neither be isolated from other terms, nor be the kind of phenomenon that can serve as a foundation" for politics (2020: 186). Once "the vulnerable", at group level, are constituted as such, we must beware of the pitfalls of paternalism and respect their capacities – actual or potential – to maintain and exercise their own power. What we confront in each case is "a constellation of vulnerability, rage, persistence, and resistance", so that vulnerability "ought not to be identified exclusively with passivity; it makes sense only in light of an embodied set of social relations, including practices of resistance". The point is to recognize "how vulnerability and resistance can work together", enabling us to "identify those sites of resistance that are opened up by vulnerability" (192).

This makes for a crucial contrast between Butler's approach and that of Evans and Reid. The latter, we saw, highlight vulnerability as someone's susceptibility to being exploited, taken advantage of by others once exposed. The allegedly innate negativity of vulnerability vis-à-vis autonomy, freedom, and control makes it seem obvious that its human bearers do their best to minimize its impact as far as possible, often rendering them easy prey to various agents' endeavour to profit from it, either by exploiting that vulnerability outright or by pretending to help protect it. To suggest, as does Butler, that vulnerability may open up "sites of resistance" is to advance a counter-approach to that taken by Evans and Reid, yet without denying the very real risks of exploitation.

What kind of resistance does Butler have in mind? The way Black Lives Matter opposes systemic racial discrimination is a case in point, as are protests against policies that criminalize immigrants and refugees. Often such protests are eminently bodily, using the body as a site of resistance, wordlessly communicating "I will not disappear [or be removed] so easily", or,

"My disappearance will leave a vibrant trace from which resistance will grow", thereby "asserting their grievability within the public and media sphere" (2020: 196). To appreciate the body's significance for resistance, it needs to be understood in its full, multidimensional context:

> The body, perhaps precisely by virtue of its boundaries, is differentiated from and exposed to a material and social world that makes its own life and action possible. When the infrastructural conditions of life are imperilled, so too is life, since life requires infrastructure, not simply as an external support, but as an immanent feature of life itself.
>
> (197f.)

The political concept of self-preservation, Butler goes on to maintain, tends to neglect the

> the preservation of the self requires the preservation of the earth, and that we are not 'in' the global environment as self-subsisting beings, but subsist only as long as the planet does. What is true for humans is true for all living creatures who require non-toxic soil and clean water for the continuation of life.
>
> (199)

In moving so swiftly from talking about the body to talking about the self, Butler brings Thomas Hobbes to mind, but does not mention him. Though worlds apart philosophically and politically, Butler and Hobbes agree that the human self is inseparable from its bodily existence, and that this fact – condition – entails that human life is not to be had without its bodily anchored vulnerability, hence mortality. Whereas for Hobbes, as we noted earlier, all activities of the self are oriented towards the reduction of vulnerability, Butler draws the opposite conclusion: "Persistence in a condition of vulnerability proves to be its own kind of strength, distinguished from one that champions strength as the achievement of invulnerability" (201). It is tempting to say that the latter view is articulated in Trump's ideal of masculinity (discussed above) and that it can be traced back to Hobbes. But this is not so: if Trump and his present-day supporters, overwhelmingly male, believe that invulnerability can ever be achieved, making sure it is something – a source of weakness – that applies only to others, and can be attacked as well as exploited there, Hobbes begs to differ – he never thought that invulnerability is within my, or anyone else's, reach and control. In this, he remained a realist about the human condition, and as such would be quick to point out the arrogance and hubris of those who, like Trump, buy into the fiction that they can overcome their own vulnerability by projecting it onto others.

What Butler recognizes more clearly than Hobbes, though, is that the question "how to view and to deal with the *conditio humana* of vulnerability" is, and will always be, a matter of culture, and hence subject to various interpretations and assessments throughout human history. In addition to *evaluating* vulnerability in specific ways, on behalf of its individual members, culture also supplies them with a specific symbolic repertoire – be it rich, be it poor, depending on social class, etc. – indispensable to tackle vulnerability whenever it is actualized in the individual's life. A key feature of the currently hegemonic neoliberal culture is that it portrays given conditions of life – vulnerability, dependency, mortality, the precariousness of relations with others, and loneliness – as "negative", since hampering our ambition for freedom and self-realization, regarded as *premised upon* our success in neutralizing the condition of vulnerability in particular (see Vetlesen 2009). Hobbes failed to recognize the historical and cultural selectivity of his own attempt at a universally valid portrait of the role played by vulnerability in human life in general and in political life in particular. And those who pretend, today, that such a view is the only natural one, or some inevitable harsh truth about "who we are", help cement that view, a product of ideology, concealing how its current hegemony suppresses alternative ways of looking upon and dealing with the conditions in question, be it previously in Western culture, or in non-Western ones still existing.

The cultural hegemony I talk about here is not a total such: there is resistance to it, as can be seen in the way women tend to differ from men in how they look upon and respond to vulnerability-dependence in various social relationships. I examined this above and shall not repeat it here. What matters now is that Butler, in the last quotations I brought, opens up a further dimension of vulnerability, one helping situate the ways in which vulnerability typically plays out inter-subjectively, between individuals – namely, when she writes that "the preservation of the self requires the preservation of the earth", and that we humans "subsist only as long as the planet does" (2020: 199). In a sense, to say this is to state the obvious in the era of the Anthropocene. Living in the twenty-first century, we all know that we are dependent on the earth, and that given the thrust of ongoing degradation and decimation of countless other-than-human species and lifeforms, we are extremely vulnerable in the face of the consequences, whereby – as Butler helps us recognize – this "we" is no monolithic such but divided along dimensions of class, race, and gender.

Yet the fact is that the other-than-human dimension of vulnerability-dependence is mentioned only in passing in her books, restricted to the sentences just quoted. Butler, that is, concentrates on humans and humans only in her account of lives deemed ungrievable and individuals (or groups) regarded as killable. The anthropocentrism of her perspective may come as a surprise, or even as a big disappointment, in being so conventionally held on to by a prominent radical thinker. Could she have included other-than-humans without undermining her critical objectives? It is tempting to say yes – but this does not answer why she excludes, say, animals from her discussion, in a situation where the exploitation of animals is such an integral part of the power structures she sets out to expose (see Crary & Gruen 2022). Her silence about the predicament of other-than-human vulnerable lives, except for the few general remarks cited, means that Butler's approach fails to critique and to overcome the "nature deficit" found in liberal political theory (Locke, Kant, Rawls, Gardiner) that I discussed in Chapter 1 (see Mulgan 2011; Vetlesen 2023).

3.7 Distrusting institutions

More important to the present discussion, though, than this nature deficit in Butler is the question I raised earlier about how deep her scepticism to modern institutions runs. To put it simply: Can modern institutions – the government in particular – be expected to protect people affected by precarity against its negative consequences? Or is such an expectation naïve? If so, does that mean that, under current conditions, institutions are a part of the problem rather than a part of the solution? But if that is the case, how can "persistence in a condition of vulnerability [prove] to be its own kind of strength", as we saw Butler contend? If the system with its major institutions is broken, how can the responsibilization of the individual that is part and parcel of the rise of resilience as we have explored it, be resisted?

In his book *The Art of Shrinking Heads: The New Servitude of the Liberated in the Era of Total Capitalism*, the French philosopher Dany-Robert Dufour writes:

> The great novelty of neoliberalism, as compared with earlier systems of domination, is that the early systems worked through institutional controls, reinforcements, and repression, whereas the new capitalism runs on deinstitutionalization. Foucault probably did not see this coming. Being fully preoccupied with studying the ways in which power takes control of life (health care, education, forms of punishment) he failed to see that a very novel domination had been gradually established from the end of the Second World War onwards. Foucault's exemplary studies of disciplinary societies in fact appeared at the moment when those societies were lapsing into decadence. They applied to an object that had already become very fragile even as he was studying it. Foucault's studies did not take into account that *the institutions they were targeting were the very apparatuses that the most aggressive faction of capitalism wished to destroy*. [...] The new capitalism was in the process of discovering and implementing much less coercive and much less costly ways of guaranteeing its own success: it

is no longer reinforcing the secondary domination which has produced submissive subjects. It is destroying institutions and putting an end to primal domination in such ways as to produce individuals who are supple, insecure, mobile and open to the market's mode and variation.

(Dufour 2008: 157; italics mine)

One may well ask whether this critical assessment of Foucault's work is valid; and even if it is, how relevant it is with respect to Butler's position. Indeed, the jury is still out on whether Foucault was genuinely fascinated by neoliberalism and some of its key thinkers (like Friedrich von Hayek and Gary Becker), seeing the potential for a new type of subjectivity; or if he remained agnostic regarding neoliberalism's political consequences, as he – for most readers – appears to be in his much-discussed lectures at the College de France in 1978–1979, posthumously published in the book *The Birth of Biopolitics* in 2008.

The exact way to interpret Foucault, and to assess Butler's debt to his work, is not the main issue, though. The important issue is the *simultaneity* of the "post-modern" turn in philosophy and the triumph of neoliberalism in politics that took place in the late 1970s and early 1980s that Dufour points out. His critical thesis is that the rise of the post-modern paradigm, especially among the radical left inspired by French philosophers such as Foucault, Baudrillard, Lyotard, and Derrida, in effect *disarmed* social and political critique. Key to this is the Nietzschean predilection for questioning – or "deconstructing" – traditional standards of validity, such as the distinction between true and false, right and wrong, exposing them as mere instruments of the powers that be. "Reason" itself was revealed – and so delegitimized – as a vehicle of power, that is, as precisely *not* the medium where the most convincing argument prevails, power can be critiqued, and falsity replaced by truth, injustice by justice (see Habermas 1987). The British sociologist Susan Stedman Jones puts the result like this, in a way consistent with Dufour's diagnosis:

The intellectual, moral and social paupacity of Thatcherism was scarcely touched by all the new sociologies – and the sphere of the social and the collective had no intellectual defenders when the political bandits arrived and sold off what remained of collective assets to the lowest bidder. Foucault's explicit nominalism underscores no collective concepts and thus logically neglects the structural: is it not thus a match for Thatcher's atomistic vision?

(Stedman Jones 2001: 224)

Whether intended or not by Foucault and other thinkers associated with post-modernism, the fact is that the "methodical destruction of all collectives" that Pierre Bourdieu (1998: 95f.) identified as the goal of neoliberalism was to prove very successful, facing modest intellectual resistance at the time it was implemented. Indeed, if regarded as sites of suppression and exploitation, why be worried about the attack launched on institutions run by the state? Shouldn't such attacks be welcomed as subversive? Perhaps the dismantling of what used to make up "the social state" in Bauman's sense would prove liberating, the entry into arrangements less stifling and oppressive – especially so in the name of the individual's self-realization and authenticity, the buzzwords shared by many leftists and neoliberals alike (see Boltanksi & Chiapello 2005)?

Forty years on, in the United States and increasingly in Europe as well, we can conclude that the dismantling of the social state, as part of the attack on "big government", has proved anything but liberating and empowering to those who historically have been most dependent on that state and its provisions. Dufour points out that "the strength of neocapitalism lies, paradoxically, in the weakness of its governments" (2008: 165), adding that "the absence of any real government, in other words of an institution whose legitimacy is of necessity unrelated to economic interests, abolishes authority and at the same time makes power invisible" (p. 166).

However, it is more accurate to speak of the *transformation* of the state and how it functions than to create the impression that the state is weakened or even about to disappear as far as the exercise of power is concerned. The still valid insight in Foucault's analysis of neoliberalism is his recognition that whereas traditional liberalism aimed to "free" the economic as well as the political and civic subject by protecting market activities vis-à-vis the state and its interventions, what is truly radical about neoliberalism is that it aims to "raise the market itself to a principle of all life or of government", as Wendy Brown (2015: 61) explains in *Undoing the Demos: Neoliberalism's Stealth Revolution*. Neoliberalism activates the state to "facilitate economic competition and growth and to economize the social", so as to "regulate society by the market" (Foucault 2008: 145), *society in its entirety*, that is, seeing to it that competition becomes the primary mode of interaction not only between economic agents but between individuals in all sorts of roles, activities, and arenas. The result is that the state is no longer in a position to give people protection against how a competition-based economy creates winners and losers; rather, the state's legitimacy is now tied to how successfully it facilitates such competition, set to "govern *for* the market, not *because* of the market" (Foucault 2008: 121). When exchange is replaced by competition on the individual-to-individual basis implied, labour is replaced by human capital, so that every person is rendered as entrepreneurial, regardless of how lacking in resources the concrete individual might be, depending on existing hierarchies of class, ethnicity, and gender. "The individual's life itself", Foucault observes, "with his [sic] relationships to private property…

with his family and household, insurance, and retirement – must make him into a sort of permanent and multiple enterprise" (p. 241). In this situation, where nothing is effectively protected against nor exempted from the logic of competition, "the market is itself true and also represents the true form of all activity", so that "those who act according to other principles are not simply irrational, but refuse 'reality'" (Brown 2015: 67).

To this day, this is truly insightful analysis. But I disagree with Foucault and Brown when they hold that the accentuated "emphasis on entrepreneurship and productivity replaces an emphasis on commodities and consumption" (Brown 2015: 65). Rather than the one replacing the other, the link between them is being strengthened insofar as the commodities you buy are perceived as telling the truth about how hard you work with yourself, and (increasingly in recent years) with your body and your looks, understood as the "human capital" you need to optimize. Indeed, the rise of resilience that we explored above should be viewed as a logical continuation of the trend that Foucault set out to describe, anticipating a change in mentality years before it came into proper effect thanks to neoliberal reforms in the guise of New Public Management. That said, I agree with Brown's claim that Foucault was "surprisingly unimaginative about the implications of the neoliberal fashioning of the subject as human capital" (78). But whereas Brown is mainly concerned with how Foucault's failure to abandon the traditional liberal notion that "*homo oeconomicus* is a man of interest" hampers his take on how neoliberalism leads to "the closing of the political universe" (citing Herbert Marcuse's thesis in *One-Dimensional Man*) (68), my concern is a different one.

Foucault's major claim that the capitalist's exploitation of the worker is being replaced by the entrepreneur's exploitation of him- or herself qua "human capital" sits well with the opening section of this chapter: Mark Fisher's account of the self-instrumentalization whereby individuals are required to display self-control and self-responsibility ("Selbstverantwortung"), relentlessly monitoring and assessing their own performance with a view to satisfying benchmarks dictated to them from whatever profession or institutions they belong to, yet internalized as were them of their own making. We saw that this cult of personal responsibility in the name of "freedom" and "individual choice" prompts the question of what happens to institutions. Do they become superfluous? Can disciplining take place without them? Do they more or less disappear?

Late Foucault to a large extent answers "yes" to the first two questions. The last one is more complicated, the point being that institutions have been transformed to facilitate the "marketization" of society in its entirety that has taken place in the last forty years. Yet the validity of Foucault's analysis does not mean that Dufour is wrong to point out how his thoroughgoing suspicion of the institutions of modern society – prison, hospital, and school

– has played a role in delegitimizing them – and this at the very time when they were being partly dismantled, partly transformed to facilitate the triumph of neoliberalism. To put it crudely: if the state is (or has been) the site of oppression, the individual is (or can become) the site of subversion and liberation. This is the notion on which there is an agreement between liberals old and new and Foucault. I therefore agree with Brown when she holds that "Foucault is clearly intrigued by the "freedom" that liberalism and neoliberalism promise" (2015: 76).

My claim is that the lowering of expectations we encountered in individuals like Sally and Eirik in Chapter 2 is illustrative of the loss of trust in relevant institutions to protect ordinary people from adversity. The very notion that institutions, or collective bodies in general, may provide such protection – be it against ongoing environmental degradation where people live – is, as we saw, conspicuously absent in Sally's and Eirik's narratives, much as it is their strategies for coping, individualized as these are – which illustrates the point. As we have seen, this mirrors the absence of trust in institutions in thinkers like Butler as well.

3.8 Nature deficit, left-oriented style

A different kind of absence is no less relevant to my discussion, however: the absence of an account of how the triumph of neoliberalism affects the natural environment, as distinct from the social one. That the domination of nature and the domination of humans go hand in hand is a crucial insight in the Frankfurt School, powerfully articulated by its founding members Max Horkheimer, Theodor W. Adorno, and Herbert Marcuse. Despite this insight into how the quest for mastery over nature is integral to capitalist modernization, the exploitation of other-than-human nature was soon to be eclipsed by the concern with how that outward-directed mastery was turned into a wholesale mastery of humans' inner nature (see Vetlesen 2015: 52-95). What truly matters to the critical theorist, then, is not the degradation of the natural world but the conflicts and pathologies occurring in the human-to-human dimension. In other words, how humans come to exploit each other, and each individual to exploit him- or herself regarded as "human capital", is what enjoys pride of place in the critique, silencing the parallel degradation inflicted upon nature.

This selectivity is not confined to a late critical theorist like Jürgen Habermas or a champion of liberal political theory like John Rawls (as I pointed out in Chapter 1). It is also, we now see, present in the work of Foucault and in contemporary left-oriented thinkers like Butler and Brown. Why isn't the ongoing degradation of the natural world a topic in its own right in their work?

The answer that first comes to mind is that their critique of the current neoliberal phase of capitalist society is overwhelmed by the sheer magnitude of the human consequences. Human suffering is almost everywhere on the rise, resulting from poverty and marginalization along divisions of class, race, gender, and geography. Natural disasters such as hurricanes, floods, and wildfires have become more frequent and help make the precarious situation of those hardest hit by the downsides of neoliberalism's reign even more difficult. Shouldn't this suffice for a critical theory these days?

I don't think so, and I have argued as much in my previous book *The Denial of Nature*, focusing on the Frankfurt School's German thinkers and keeping French ones like Foucault out of the picture, as well as American ones such as Butler. Yet we need to realize that what is a conspicuous absence in the former is no less so in the latter.

But why, more precisely, is this significant? Why not settle for the *de facto* division of intellectual labour we witness today, where critical social theory concentrates on exactly that, critiquing dominant social arrangements with regard to patterns of repression and injustice, leaving it to environmental philosophy to concentrate on this society's relentless exploitation of the natural world?

This will not do, for the simple reason recognized in early Critical Theory's fundamental insight: the two processes of domination and exploitation go hand in hand; indeed, they are inextricably intertwined, causally and dynamically so. This being the case, a programmatic decision to keep them apart as were they separate in how they operate is not only inadequate, but risks concealing the interplay between the two dimensions of exploitation at a historical moment when clarifying their interconnectedness is more important than ever. Exhaustion of oneself as human capital, taking the form of burnout, and exhaustion of outer nature as nature capital, taking the form of overshoot, are two aspects of the same system.

Enough theory. It is time to turn to how the double exploitation I have talked about is experienced by some of the people who are most affected by it, in the contemporary American society that is the backdrop of Butler's and Brown's recent books. The story I shall quote is one told to the sociologist Arlie Russell Hochschild by members of a family in Louisiana:

> Now retired from AT&T and working as a substitute teacher, Dicky had been a young boy in the 1950s. "I was riding my palomino horse, Ted", he recalls. "Normally Ted cleared ditches five feet across just fine. But this time the horse fell back into the water and sank down. He tried to climb up but couldn't. We tried to pull his reins, but couldn't get him up. Finally my uncle hauled him out with a tractor. But when Ted finally scrambled back out, he was coated all over with a strange film. I hosed him off but that only hardened the film on him. It was like a terrible glued-on wet suit. It was like rubber. The vet tried but couldn't save him, and Ted died two days later". The ditch was downstream from a Firestone polymers

plant. Dicky was heartbroken at the time, and it still shows as he tells the story. But Janice recalls the episode slightly differently. It saddens her too, but she doesn't allow her sadness to interfere with her loyalty to industry. She shakes her head as if to say, we sure did put up with a lot of things back then, but let's not linger on too much bad news, the way environmentalists do.

(Hochschild 2016: 163f.)

Doing research for her book *Strangers in Their Own Land: Anger and Mourning on the American Right*, Hochschild stayed for long periods with families such as this one in Louisiana, getting to know them well, allowing them to open up about thoughts and feelings they usually kept inside. A professor at the University of Berkeley, California, just like Butler and Brown, Hochschild comes from a truly different background than the mostly working-class people she talked to, many of whom support the Tea Party movement and vote for Trump. "Those I came to know", writes Hochschild,

> developed a visceral hate for the ally of the 'enemy' cutters in the line [people exempt from waiting for their proper turn, thanks to affirmative action and other so-called progressive policies meant for various minorities] – the federal government. They hated other people for needing it. They rejected their own need of it – even to help clean up the pollution in their backyard.
>
> (2016: 151)

For Janice and others like her, with the same history in the region, Hochschild notes that "her distrust has gone the full cascade: from president to the redistributive function of the federal government to nearly all government functions – including that of cleaning up the environment" (178).

What struck Hochschild is reminiscent of what struck Lertzman in Green Bay and Norgaard in Bygdaby: the combination of visible and as such undeniable environmental degradation in one's home place, and a near total silence about it. When Hochschild brings up the environment, the response is:

> Pollution? I don't talk about it much with friends. This whole town operates off of oil. So I could be talking to two moms whose husbands work in the plants. They think government regulation will hurt jobs, or stop new plants from coming in. You don't want to *remind* them of dangers. Or make them think you're *blaming them* for the work they do. It's too close to home.
>
> (176)

too close not only physically, but psychologically and emotionally. Hochschild sums up the position taken by her respondents: "A clean environment? Sadly, we can't have it." In Jackie's words: "Pollution is the sacrifice we make for capitalism" (179).

The bottom line, then, is this: "It's worth it". What can't be changed must be adapted to. To be sure, there are real problems, with valid reasons for concern, but it is better to focus on the positives, such as the fact that even though the factory brings pollution to the river and the air, it also provides a salary. The factory is a double-edged entity, recalling Lertzman's talk about "industrial ambivalence" when listening to Sally's story in Green Bay: on the one hand, the factory degrades the environment, on the other hand, it makes it possible to pay the bills – to stay in one's rural home area, even though it's going from bad to worse as far as the environment is concerned.

If pollution is a sacrifice that is worth it, that must also include the horse Ted mentioned above. Having lost a much-loved horse this way is, admittedly, truly heartbreaking. The impression I get – and I improvise since Hochschild does not expand – is that based on such an experience, acquiring a new horse is out of the question. Other things are too, like fishing in the river, going for a swim in the lake, or encouraging one's children or grandchildren to try and catch frogs or, when old enough, to take part in hunting for game. So many activities that were a significant part of growing up in the countryside, far from the noise and the adventures on offer in the city, being something it's okay to seek out in weekends for the fun of it, knowing it's always possible to retreat to the calm of a house with a big garden, to the open fields, to the doorstep from which the stars can be seen when night falls, and the owl can be heard, if only I concentrate properly, sharpening my hearing and trusting that if I am patient, its cry in the night will reach me.

There is no need to go on. The natural world that used to be part and parcel of the social one, a two-in-one neither experientially nor existentially to be separated, is no more, and the social lifeworld suffers as a consequence. Life is impoverished, so many versions and variations of what used to make it rich, manifold, enthralling, are in decline or have altogether disappeared. It's not only the wild animals, the birds, the fish, and the game that cease to show up, to be seen and heard, as part of the loss of the places – habitats – that used to be their home too, not only mine. It's also domestic animals like the horse, and having witnessed its death from pollution not only signals that getting a new one would be too risky, or associated memory-wise with too much pain – you never forget the scene, here at this place – it is also a warning to think twice about acquiring a dog of a kind that favours roaming outdoors. With so many elements – creatures, lifeforms – of the natural world being degraded or simply vanishing, the result is a new sort of absence, of silence, nowadays more multi-caused than Rachel

Carson's pesticide-induced *Silent Spring* in the early 1960s; a silence not quite right, conveying that something is wrong, amiss, hence an uncanny silence, causing discomfort and anxiety.

The result is the need to adjust to a new normal that is all about the losses inflicted and accumulated here, at home, eminently tangible at that. Only it doesn't feel as homely as it used to; its taken-for-granted trustworthiness has become precarious, and it is in danger of being lost, with nothing to replace it. Writ large: if the natural world, since time immemorial, represented, and sustained something stable in how spring follows winter, fall follows summer, each season eminently recognizable in the changes brought and the manifestations of life that go with them – birdsong not only greeting spring, but *being* spring – such season-based stability can no longer be taken for granted. It is being disrupted before our eyes – beyond recognition and, perhaps, beyond reparation.

Bibliography

Adorno, Theodor W., 1998. *Critical Models*. New York: Columbia University Press.
Alford, C. Fred, 1988. *Narcissism. Socrates, the Frankfurt School, and Psychoanalytic Theory*. New Haven: Yale University Press.
Alford, C. Fred, 1989. *Melanie Klein and Critical Social Theory*. New Haven: Yale University Press.
Alford, C. Fred, 1991. *The Self in Social Theory*. New Haven: Yale University Press.
Alford, C. Fred, 1994. *Group Psychology and Political Theory*. New Haven: Yale University Press.
Alford, C. Fred, 1997. *What Evil Means to Us*. Ithaca: Cornell University Press.
Alford, C. Fred, 2005. *Rethinking Freedom*. New York: Palgrave Macmillan.
Bauman, Zygmunt, 1995. *Postmodern Ethics*. Oxford: Blackwell.
Bauman, Zygmunt, 2008. *Does Ethics Have a Chance in a World of Consumers?* Cambridge, MA: Harvard University Press.
Beck, Ulrich, 1986. *Risikogesellschaft. Der Weg in eine zweite Moderne*. Frankfurt/M.: Suhrkamp.
Benjamin, Jessica, 1988. *The Bonds of Love*. New York: Pantheon.
Boltanksi, Luc, and Eve Chiapello, 2005. *The New Spirit of Capitalism*. New York: Verso.
Bourdieu, Pierre, 1998. *Acts of Resistance*. Cambridge: Polity Press.
Brown, Wendy, 2015. *Undoing the Demos. Neoliberalism's Stealth Revolution*. New York: Zone Books.
Brundtland, Gro Harlem, 1987. *Our Common Future*. Oxford: Oxford University Press.
Butler, Judith, 2004. *Precarious Life*. New York: Verso.
Butler, Judith, 2009. *Frames of War. When is Life Grievable?* New York: Verso.
Butler, Judith, 2020. *The Force of Nonviolence. An Ethico-Political Bind*. New York: Verso.
Camassa, Manuel, 2024. *On the Power and Limits of Empathy*. New York: Palgrave Macmillan.
Carson, Rachel, 1962. *Silent Spring*. Boston: Houghton Mifflin.

Ceballos, Gerardo, Paul Ehrlich, and Rodolfo Dirzo, 2017. "Biological annihilation via the ongoing sixth mass extinction signaled by vertebrate population losses and declines", *Proceedings of the National Academy of Sciences of the United States of America* 114 (30): E6089–E6096.
Crary, Alice, and Lori Gruen, 2022. *Animal Crisis. A New Critical Theory*. Cambridge: Polity Press.
De Beauvoir, Simone, 1956 [1949]. *The Second Sex*. New York: Harper Collins.
Dufour, Dany-Robert, 2008. *The Art of Shrinking Heads*. Cambridge: Polity Press.
Evans, Brad, and Julian Reid, 2014. *Resilient Life*. Cambridge: Polity Press.
Fisher, Mark, 2009. *Capitalist Realism*. Winchester: Zero Books.
Foucault, Michel, 2008. *The Birth of Biopolitics. Lectures at the College de France 1978-1979*. New York: Palgrave Macmillan.
Gilligan, Carol, 1982. *In a Different Voice*. Cambridge, MA: Harvard University Press.
Graefe, Stephanie, 2019. *Resilienz im Krisenkapitalismus. Wider das Lob der Anpassungsfähigkeit*. Bielefeld: Transcript Verlag.
Habermas, Jürgen, 1987. *The Philosophical Discourse of Postmodernity*. Cambridge: Polity Press.
Hansen, James, 2009. *Storms of My Grandchildren*. New York: Bloomsbury.
Hegel, Georg F. W., 1991 [1821]. *Elements of the Philosophy of Right*. Cambridge: Cambridge University Press.
Hochschild, Arlie Russell, 2016. *Strangers in Their Own Land. Anger and Mourning on the American Right*. New York: The New Press.
Klein, Melanie, 1989. *Love, Guilt, and Reparation*. London: Virago.
Klein, Naomi, 2014. *The Shock Doctrine*. New York: Penguin.
Lertzman, Renee, 2015. *Environmental Melancholia. Psychoanalytic Dimensions of Engagement*. New York: Routledge.
Levinas, Emmanuel, 1985. *Ethics and Infinity*. Pittsburgh, PA: Duquesne Press.
Levinas, Emmanuel, 1991. *Totality and Infinity*. Dordrecht: Kluwer.
Levinas, Emmanuel, 1993. *Otherwise than Being or Beyond Essence*. Dordrecht: Kluwer.
Løgstrup, Knud E., 1956. *Den etiske fordring*. København: Gyldendal.
Mirowski, Philip, 2014. *Never Let a Serious Crisis Go to Waste*. New York: Verso.
Mulgan, Tim, 2011. *Ethics for a Broken World*. New York: Routledge.
Nietzsche, Friedrich, 1970 [1889]. *Twilight of the Idols*. New York: Viking.
Norgaard, Kari Marie, 2011. *Living in Denial. Climate Change, Emotions, and Everyday Life*. Cambridge, MA: MIT Press.
Norris, Christopher, 1992. *Uncritical Theory. Postmodernism, Intellectuals, and the Gulf War*. London: Lawrence and Wishart.
Scheff, Thomas, 2006. "Aggression, hypermasculine emotions and relations: the Silence/violence pattern", in *Irish Journal of Sociology*, vol. 15 (1), 24–40.
Schopenhauer, Arthur, 1986. "Über die Grundlage der Moral», in *Sämtliche Werke*, vol. 3. Frankfurt/M.: Suhrkamp.
Skjerven, Astrid, et al, eds., 2025. *Mediating Sustainability in the Consumer Culture*. London: Routledge.
Stedman Jones, Susan, 2001. *Durkheim Reconsidered*. Cambridge: Polity Press.
Stoknes, Per Espen, 2015. *What We Think About When We Try Not to Think About Global Warming*. White River Junction: Chelsea Green Publishing.
Trump, Mary L., 2020. *Too Much and Never Enough. How My Family Created the world's Most Dangerous Man*. London: Simon & Schuster.

Vetlesen, Arne Johan, 1994. *Perception, Empathy, and Judgment*. University Park, PA: Penn State Press.
Vetlesen, Arne Johan, and Harald Jodalen, eds., 1997. *Closeness. An Ethics*. Oslo/Boston: Scandinavian University Press.
Vetlesen, Arne Johan, 2005. *Evil and Human Agency. Understanding Collective Evildoing*. Cambridge: Cambridge University Press.
Vetlesen, Arne Johan, 2009. *A Philosophy of Pain*. London: Reaktion Books.
Vetlesen, Arne Johan, 2015. *The Denial of Nature. Environmental Philosophy in the Era of Global Capitalism*. London: Routledge.
Vetlesen, Arne Johan, 2023. *Animal Lives and Why They Matter*. London: Routledge.
Vetlesen, Arne Johan, 2025. "*Sustainability and its Discontents: Recovering the Sources of Value*", in Astrid Skjerven, et al., eds., 13–30.
Williams, Robert, 1997. *Hegel's Ethics of Recognition*. Berkeley, CA: University of California Press.
Wolin, Sheldon, 2004. *Politics and Vision*. Princeton: Princeton University Press.

4 Resistance

Activism, whistleblowing, and leadership

4.1 Civil disobedience: the "emergency break"

In *Requiem for a Species*, Clive Hamilton observes that "no government seems willing to do what is needed to tackle climate change, which is to speak to their citizens as adults, explaining that we face an emergency, one that can be met only at some cost and inconvenience" (2010: 173). Writing in the aftermath of the failed Copenhagen conference in December 2009, Hamilton contends that this conference "was the last hope for humanity to pull back from the abyss" (xi).

Fifteen years later, how does Hamilton's statement hold up? Has his grim conclusion proved right? What difference does it make that the 2015 conference in Paris was hailed as a success? Or do we have to admit that, despite that breakthrough, things have gone from bad to worse?

There is no doubt that Hamilton's verdict is even more valid today than it was in 2010, namely that "the national and international political institutions that must bring about the changes are too slow, too compromised and too dominated by old thinking to mandate the energy revolution we must have to guarantee our survival" (173). If the political institutions fail in the most important task entrusted to them by citizens, in securing our survival, what shall we do?

Hamilton's answer is worth quoting in full:

> We all value and benefit from a law-abiding society. Yet at times like these we have a higher duty and are no longer bound to submit to the laws that protect those who continue to pollute the atmosphere in a way that threatens to destroy the habitability of the Earth. When just laws are used to protect unjust behaviour our obligation is to uphold the laws is diminished. [...] Those who engage in civil disobedience are usually the most law-abiding citizens – those who have most regard for the social interest and the keenest understanding of the democratic process. [...] Only by acting, and acting ethically, can we redeem our humanity.
>
> (2010: 225f.)

DOI: 10.4324/9781003660743-5

Breaking the law, then, can be the ethically right thing to do, indeed what we are ethically *demanded* to do, and be it in well-functioning liberal democracies like Australia (where Hamilton lives) or my native Norway. This demand takes primacy over ordinary political procedures as well as legal prescriptions. It comes to the fore, and becomes necessary, as a response to the failure of politics and law alike; it is an ethical demand predicated on this double failure. As such it means resorting to an action of emergency in the face of emergency. It shouldn't be necessary to do so, but it is.

Civil disobedience has been recognized as the "emergency break" of a democratic society by philosophers such as Rawls and Habermas. The most influential discussion is to be found in Rawls's seminal book *A Theory of Justice* (1971), where he defines civil disobedience as

> a public, nonviolent, conscientious yet political act contrary to law usually done with the aim of bringing about a change in the law or policies of the government. By acting in this way one addresses the sense of justice of the majority of the community.
>
> (1971: 364)

In justifying civil disobedience, Rawls goes on, "one does not appeal to principles of personal morality or to religious doctrines"; nor can such obedience be "grounded solely on group or self-interest. Instead one invokes the commonly shared conception of justice that underlies the social order" (365). Civil disobedience is a public act, addressed to public principles, and carried out in public. Hence it is comparable to public speech, being a form of address, one demanding and deserving a reasonable response. Rawls stresses that civil disobedience is, and needs to be, nonviolent, incompatible with "violent acts likely to injure and to hurt", since this would interfere with the civil liberties of others (366). Civil disobedience qualifies as a "last resort" when, and only when, "the normal appeals to the political majority have already been made in good faith and they have failed. The legal means of redress have proved to no avail" (373).

These being the criteria of civil disobedience set out by Rawls, the question is: Do they fit with the sort of actions carried out by today's climate activists?

As it happens, I have been called as an expert witness in two such cases before the court in Oslo. In the first case, 17 activists from Extinction Rebellion (XR) were indicted; in the second, two activists from Just Stop Oil. To concentrate on the case against the XR activists, the charges brought referred to "causing disturbance to the public order", be it in the form of disrupting a debate in the parliament, blocking car traffic in central Oslo during rush hour, and occupying the reception hall in the Department of Energy.

For a start, the activists consider their organized protests as a last resort, having tried all ordinary means to have their arguments about the need to stop Norwegian oil production heard, such as writing in the papers, inviting politicians to debate their demands, and being vocal members of various environmental groups, such as WWF and Greenpeace. These efforts have all been to no avail; in fact, instead of reducing oil production, various governments have given the green light to expand the number of sites for drilling in the Barents Sea, aiming to be operative until 2070. The activists, most of whom are in their twenties, regard this policy as in clear violation of the Paris Agreement signed by Norway in 2015.

Second, the issue at stake for the activists is bigger than each one of them as a private person. Their concern has to do with the wider community, wider in the sense of giving voice to those who are not able to promote their interests and rights, such as animals and other non-human creatures, and future generations. Third, their acts of protest are public and out in the open, making sure that the public will notice them and be encouraged to take a stand on the issues involved. Given the media coverage that goes with the protests being public, the authorities have to respond. The public demands a response that is reasonable, taking seriously the issues raised and the claims made by the protesters. Fourth, the activists fully recognize that what they are doing is illegal according to existing laws. With open eyes they resort to illegal acts to articulate their view on an issue they cannot, existentially and morally, be silent about, leaving it to others to sort it out. Fifth, the actions they engage in are non-violent, lending them a moral authority, even should they be met with violence, in keeping with Mahatma Gandhi's principles: no one is to be hurt or put into jeopardy.

This shows that the Extinctions Rebellion activists carry out their acts of protest in a manner compatible with the characteristics of civil disobedience listed by Rawls. Greta Thunberg's first act of protest is another case in point: in August 2018, Thunberg, fifteen years old at the time, dropped school to sit down, alone, on the pavement in front of the Swedish parliament in Stockholm, carrying a poster where she had written "Skolstreik för klimatet" ("school strike for the climate"). Her action satisfies all of Rawls's criteria. Its novelty consists in the directness of the way she appeals to adults in general, holders of political power in particular: they have failed in their responsibility to make sure that her generation, and future ones, take over a sustainable and habitable planet. In protesting the way she does, as a child and with no one telling her to do so, she feels obligated to do something she should not need to do: the betrayal of adults, acting – or non-acting – in a manner totally at odds with their responsibilities and power, is what forces her to protest in public.

Even though both Extinction Rebellion activists and Thunberg carry out their protests in accordance with Rawls's criteria, they also show up some limitations to his approach. The first is to do with the distinction between a majority and a minority that is so important to him:

> By engaging in civil disobedience a minority forces the majority to consider whether it wishes to have its actions construed in this way, or whether, in view of the common sense of justice, it wishes to acknowledge the legitimate claims of the minority.
>
> (1971: 366)

In his discussion Rawls admits that his theory "presupposes that the majority has a sense of justice", and that this may appear problematic. His response is that "I have assumed throughout that we have to do with a nearly just society" (386).

I don't find this reply very convincing. But the more important question that Rawls does not mention at all, is this: why does he build the assumption that those engaging in civil disobedience are "a minority" into his very definition? Is it for conceptual reasons? Or for empirical ones? Does history tell us that people resorting to civil disobedience are always a minority? Or that the views they articulate represent those of a minority?

To be sure, it seems reasonable to say that those who actively engage in civil disobedience form a smaller group than those they are appealing to. Even so, to act so as to appeal *to* a majority is not to be confused with *representing* the views and interests of a minority. This is illustrated in the case of Thunberg. She is protesting against current climate policies on behalf of *everybody* affected, and likely to be so. In particular, as a child she is making her appeal on behalf of a generation too young to vote. Hence her appeal goes beyond the here and now, being concerned with the future in a temporal perspective not limited to her life when herself an adult but including her potential children, and with that future generations. To frame the climate issue as one crucially involving the future is to shift the emphasis from those living today to those not yet born, and so unable to be party to agreements made today, with the reciprocity and give-and-take that such arrangements imply, typically taken for granted in liberal contract theory, as I pointed out in Chapter 1. Future generations are one-sidedly dependent on the efforts of present individuals to speak on their behalf.

We need to recognize that as far as the consequences of past and current (and so cumulative) consequences of climate change are concerned, the 9 billion people living today in fact form a minority; we do not exhaust the category of "those affected". It is arbitrary and at odds with the facts to restrict the universe of those affected, hence to be taken into account, to those who happen to live today. As a matter of principle, and in keeping with the intrinsic value of human persons (Kant), we are obliged to respect the worth of people who will live after us no less than our contemporaries.

Giving primacy to the present and those being part of it would mean to discriminate against those who will succeed us; it would be in breach of the principle of the equal dignity of all humans. This recalls the conflict between discounting and the equal worth of present and future generations discussed in Chapter 1.

Despite its philosophical credentials, this way of discussing the issue is excessively, and unnecessarily, abstract. What is important with regard to civil disobedience is the eminently concrete fact that present people prioritize their (our) perceived interests in such a manner as to *endanger*, rather than secure, the conditions of life for those who come after us. For this we are indisputably responsible, the more so, the more knowledgeable we are about the fact that this is indeed what we do.

Does Rawls say anything that sheds light on this? He states that "the injustice one protests is a clear violation of the liberties of equal citizenship, or of equality of opportunity" (1971: 375). Here, the limit of Rawls's theory that I alluded is not to do with the principles as such, but his presupposition that the persons to whom the principles apply are *presently living* persons, possessing equal rights and equal dignity, and being able to interact with each other so as to reciprocate the respect this equality commands.

The possibility of such interaction among contemporaries is precisely what is lacking in the future-oriented perspective introduced above by way of Thunberg's protest. But the limit I am concerned with is not simply that Rawls, like almost all philosophers before him, fails to take the future into account. The more immanent problem in his theory of justice is that, in presupposing what he calls "favourable conditions", he takes it for granted that *basic liberties will not conflict with basic needs*. In keeping with this programmatic optimism of his, Rawls assumes that each new generation would be better off than the previous one (1971: 542ff.). Given his trust in the stability of the present political institutions, and more tacitly, his trust in the stability – intactness – of the natural world and its "ecosystem services" to human society, it simply did not occur to Rawls that present people – especially in affluent societies like the United States – *enact their basic liberties* in such a way as to endanger the equal liberties as well as basic needs of the generation(s) to succeed them, including their own children and grandchildren.

In holding that "liberty can be restricted only for the sake of liberty" (1971: 244), focusing on liberty of participation, thought and conscience, Rawls keeps the *material* and nature-related consequences of the practices of production, distribution, and consumption of goods, bought and possessed as part of one's liberty to do whatever one wishes with one's money, outside his framework. It is as if the enactment of liberty has no "outside", no embeddedness in and dependence upon the natural world, only an "inside" to do with the interactions with other people. In effect, if not by intention, then, the ways in which human society impacts nonhuman nature are ethically neutral; the very distinction between right and wrong is reserved for inter-human affairs. At one point Rawls acknowledges this, stating that "no account is given of right conduct in regard to animals and the rest of nature", adding that the issues raised with regard to the latter "are outside the scope of the theory of justice", and leaving it at that (1971: 512). This being so, Rawls pays no attention to the ecological consequences of the level and pattern of consumption enjoyed by people (most of them, not all) in an affluent society. This is a society where advertising plays a crucial role in shaping people's perception of themselves as either winners or losers, and of their sense of entitlement and claim to social status, mediated through their behaviour as consumers. To his credit, Rawls acknowledges the importance of self-respect, considering it a primary good, yet takes no interest in how the need to attain it is deftly exploited in marketing, channelling it to the products said to help enhance it. The *material* basis of gaining self-respect is downplayed in such a manner as to portray its attainment as an almost exclusively social accomplishment.

What about Rawls's famous device, the veil of ignorance (1971: 136ff.)? Even if we grant this notion, meant to make sure that nobody knows what position they would occupy within a society they would deem just, we now recognize the huge difference it makes *when* and *where* people live. To be sure, Rawls holds that from a moral point of view, all persons have equal worth; time (epoch) and place are considered "contingent" factors, devoid of normative significance. But today we know that these factors – when and in what part of the world you are born – help determine precisely what Rawls the liberal philosopher wanted to minimize: inequality in degree of freedom and in the prospect of having basic needs met. The optimism that Rawls was committed to led him to neglect discussing population growth as well as environmental problems. That humans in the future – and not necessarily in the distant future – in increasing numbers and parts of the world would live under conditions of *deterioration* rather than persistent progress, and that they would do so partly as a *result* of progress capitalist, growth-driven and consumption-intensive style, is a scenario conspicuously absent in Rawls's work.

Given these criticisms, what would Rawls think about the following proposal for a moral principle: All human beings have the same right to live on an ecologically habitable planet? A principle with the implication that the planet's life-sustaining goods are something to which all persons are equally entitled, regardless of time and place, that is to say, regardless of generation and nationality?

On a benevolent reading of Rawls, there is nothing in what he says that prevents him from endorsing this principle. In other words, he is not committed to excluding anyone, on the basis of when they are born and where they live, from the equal right to the goods implied, since such exclusion would be arbitrary, to do with contingent factors, and so lacking in moral significance.

Yet the limits of his theory pointed out above do not disappear. Whether Rawls's theory acknowledges it or not, the factors time and place *do* make a huge difference, threatening the equality of worth Rawls sets out to safeguard across generations and locations, opening a gulf between how the world is and how it ideally ought to be. To expand the scope of ignorance about time and place as I have suggested, means to be forced to take the natural world – the environment, ecology – into account. To do so means to face the fact that due to overexploitation and overconsumption, driven in large part by how the liberties enjoyed by the most affluent are exercised, the non-human substratum on which Rawls's human-centered basic liberties as well as basic needs are dependent – since human life itself is – is no longer intact, but is being degraded, causing hardship and scarcity where Rawls takes the persistence of "favourable conditions" for granted. The upshot is absolutely crucial: a political theory that defines nature and humans' presence in it and dependence on it to be outside its scope, risks being irrelevant in today's world. The material, nature-based presuppositions of a "well-ordered" and fair society, meaning the *factual conditions* of human life on Earth, are no less important than the "ideal" ones that Rawls deliberately makes the focus of his theory. Indeed, for lack of the intactness of the former, the principle of liberty that he gives pride of place is without grounding in the material world. For lack of food, or water, or a habitable place to live, my liberty is purely ideal: I cannot flourish as a free human being if the natural world I am part of and the other-than-human creatures with whom I share the planet are not also flourishing.

But is this criticism of Rawls, and by implication the tradition of Kant-inspired contract theory he promotes, fair? Can he be blamed for not foreseeing the ecological crisis of our time when he wrote his book over 50 years ago? For a start, there were a number of warnings issued in the 1960s, Rachel Carson's *Silent Spring* (1962) being a prominent case in point. Yet there is no trace of ecological concern in Rawls's major work. In his second chance of paying attention to the crisis, his book *Political Liberalism*, published in 1993, he also fails to do so, even though the severity of the crisis was widely recognized by then.

In the 30 pages devoted to civil disobedience in *A Theory of Justice*, not a single historical case is mentioned. Rawls's determination to discuss the topic in purely abstract terms is true to his Kant-inspired philosophical approach, stressing principles and formal procedures. Even though he never says so, it is inevitable that the Civil Rights movement led by Martin Luther King Jr. forms the backdrop for Rawls's position, conceived in the late 1960s. The acts of civil disobedience carried out by King and his supporters must have struck Rawls as in tune with his model where, as we have seen, a minority takes to the streets to appeal to a majority.

A feature I find striking in Rawls's discussion is his inclination to highlight the possibility of reaching consensus, that is, of the minority being heard, over conflict. This may, once again, attest to his general optimism about progress. Yet a more critical way of putting it is to say that it, once again, demonstrates his aversion to conflict. This aversion is apparent in his explicitly stated presuppositions: that "the majority has a sense of justice" and that "we have to do with a nearly just society" (1971: 386). But isn't there something counter-intuitive about these presuppositions, especially the second one? And what exactly does "nearly just" mean – for whom, in what circumstances? When people resort to civil disobedience, don't they typically do so because of what they perceive, and first-hand experience, as *unjust* and unfair, as a reality revealing the injustice of current political arrangements and power structures?

There is but one exception from Rawls's tendency to highlight the potential for consensus over conflict, and it comes at the very end of his discussion:

> Yet if justified civil disobedience seems to threaten civic concord, the responsibility falls not upon those who protest but upon those whose abuse of authority and power justifies such opposition. For to employ the coercive apparatus of the state in order to maintain manifestly unjust institutions is itself a form of illegitimate force that men [sic] in due course have a right to resist.
>
> (1971: 391)

This, then, is as far as Rawls is willing to go when it comes to "a right to resist" institutions that are "manifestly unjust". What happens when such resistance is mobilized? How will the institutions in question, being unjust, respond? Does it suffice to state how they *should* respond – this being what Rawls does, content to do so within the framework of his ideal theory, careful not to dirty his hands by discussing a single historical example? What if the institutions in question respond by hitting back at those engaging in resistance? Or must we conclude that Rawls's theory only applies to cases where the response is in keeping with his principles of justice?

4.2 Taking the climate crisis to court

We need to go beyond the Rawlsian framework to appreciate the distinct characteristics of civil disobedience as carried out by today's climate activists. My testimony in the two court cases mentioned above is instructive here.

The essence of civil disobedience as I defined it in court is *moral*: Actions of protest that according to existing law are illegal, target practices (the extraction of oil, coal, and gas) that are currently regarded as legal, yet are considered illegitimate by the activists. The currently illegal (punishable) actions raise questions about the legitimacy of the practices in question by pointing out that they are irresponsible, unjust, and dangerous for future generations. By this reasoning, the activists adopt a civic role of advocacy: they represent parties that cannot represent themselves here and now, yet whose moral rights and claims merit being respected no less than those of presently living persons. To continue business as usual in the form of economic growth and the level of material consumption practiced by the present generation(s) is not a human right, but means stealing from future generations, violating their rights. We are about to leave the planet to our successors in a worse shape than when we took it over. We have damaged it on our watch. Those who are responsible for this have the knowledge, power and authority to see to it that the damage not become more harmful than it already is, not only to humans, but to other-than-human creatures, lifeforms, and ecosystems as well. Indeed, future human lives can only be secured by securing the latter.

By presenting their case in this manner, I argue in the spirit of Hamilton's statement cited above: "When just laws are used to protect unjust behaviour our obligation to uphold the laws is diminished" (2010: 225). In fact, an additional claim of his is even more to the point in connection with my testimony, namely that "our obligation to our fellow humans and the wider natural world entitle us to break laws that protect those who continue to pollute the atmosphere in a way that threatens our survival" (226). The stakes could not be higher.

The crux of the matter is the way in which the distinction between the legal and the illegal is understood in existing law when applied to climate change. What is the problem? It is that what counts as legal practices (tied to fossil fuels in particular) are allowed to continue along a path that we know (scientifically and empirically) will be catastrophic in terms of global warming, species extinction, and degradation of habitat. At the same time, actions that seek to *prevent* or halt our course locked to this trajectory count as illegal. To put it concretely, you may – legally – be rewarded by becoming a billionaire from investing in fossil fuel projects that cause irreversible damage to the climate and the environment, whereas if you by nonviolent means protest to stop this damage from happening, you will be fined and possibly go to prison.

Is this really the way justice, the rule of law, is supposed to work in a democracy? Or should we say that the way the distinction between what is legal and what is illegal operates here, amounts to an outright *perversion of justice*, turning the moral core of the distinction on which the institutions of law are based on its head? What happens to *legitimacy*?

In the two cases I took part in the accused activists were found guilty. Some were convicted to fines, some to a prison sentence. The Norwegian authorities were acquitted on all counts. Specifically, the verdict means that the 67 per cent state-owned oil and gas company Equinor was given the green light to implement their plans for expanding their extraction of fossil fuels for decades to come, including searching for new sites to put into production. Most experts don't think this is in keeping with the obligations to keep temperatures below 1, 5, or even 2, 0 degrees Celsius that Norway has signed up to in Paris in 2015.

That doesn't matter, though. What matters when these cases are put to the court is that the acts of civil disobedience carried out in these protests are to be considered illegal according to existing law. Their being criminal in this de facto sense renders those who took part liable to punishment. The activists' declared *reasons* for acting as they do, breaking the law, is not as such given any weight when the court reaches its verdict; the law does not engage in evaluation of motives and intentions, only in whether the principle embodied in the law is violated or not. In this way, the entire *moral* dimension so crucial to the activists and their self-understanding and manner of justification is kept at arm's length by the judges in these cases.

But, one hastens to add, the involved parties knew as much beforehand, or they should do: civil *dis*obedience is about breaking existing laws, or at least risking being considered to break them; or if not unambiguously breaking them, then *challenging* them, testing them in what is perceived as novel circumstances, different from those in which, and for which, they were originally conceived. In this way, the actions we talk about function as a *corrective* to existing laws and how they are interpreted and practiced. As a result of this confrontation, the laws may come up short; they may be seen to have deficiencies, blind spots, and limits not previously recognized. Whereas existing laws are inevitably a product of the past, acts of civil disobedience typically seek to put a flashlight to urgent matters of the present with a view to their implications for the future. Actions addressing climate change – itself an eminently dynamic phenomenon, complex and multidimensional – are a prime case in point. The dynamism of this case, of the reality it points to and its trajectory, challenges the institutions of law to *respond accordingly*: by exhibiting capacity to change, to revise and reorient itself so as to be up to the novel developments at hand, seeing to it that the rights of the citizens are being upheld and protected, now and in the foreseeable future.

The Oslo court's verdict failed the test. It failed spectacularly, demonstrating rigidity and conservatism when a dynamic response was called for by the nature and urgency of the case brought before it and the reasoning it was based upon – a reasoning whose core is moral and concerned with the *legitimacy* of existing laws, not narrowly with their de facto legality.

Some readers may find this judgment of mine too harsh. Is it realistic to expect the court to be future-oriented rather than tied to the past? After all, it may be objected, the practice of law must be based in the past: these are the laws and what they determine in terms of legal versus illegal acts; as such they cannot be modified in the courtroom, as it were, based on arguments – however valid – about the need to modify them so as to make them appropriate to, say, climate change and the challenges it poses for the institutions of law. Setting out to modify laws when arguing a specific case in the courtroom, in real time, is like putting the cart before the horse.

This objection may sound valid, generally speaking. But it does not hold in the case of Norwegian law as of today. In 2014, the Norwegian parliament, by a large majority vote, accepted that a new paragraph, § 112, be incorporated into the constitution. The paragraph reads like this:

> Everybody has the right to an environment that secures health, and to a nature where the capacity for production and diversity are protected. Natural resources shall be distributed in accordance with a broad and future-oriented viewpoint whereby this right is safeguarded also for future generations. The citizens have the right to be kept informed about the condition of their natural environment, and about the consequences of implemented as well as planned incursions into nature, enabling them to safeguard the abovementioned right. Federal (state) authorities shall implement measures to uphold these principles.

In 2016, Greenpeace along with the youth wing of Friends of the Earth Norway took the Norwegian state to court for having violated this paragraph when granting Equinor permission to search for more oil in the North Sea. The case went all the way to the Supreme Court, whose verdict in December 2021 was as follows: the Norwegian state has not violated § 112 in granting Equinor said permission; moreover, the state is only liable for emissions generated within Norwegian territory and affecting its inhabitants (in fact, more than 90 per cent of the emissions occur in the countries to which Norwegian oil is being exported). The paragraph is thus not to be interpreted as a right to which the citizens are entitled, this being the argument made by the environmental organizations; and the state's explicitly stated duty to preserve the intactness of nature for the sake of our successors, the generations to come, carries no weight in the judgment. The verdict means that, according to the Supreme Court, the institutions of law are *not* in a position to challenge, let alone overturn and set aside the climate policies of the authorities in general and the government's petroleum policies in particular. In effect, then, the court's decision gives the authorities complete freedom to open new sites for oil production within Norwegian territory. The duty to consider the consequences of continued fossil fuels extraction is pushed into the future.

The verdict wasn't only a big disappointment to those who initiated the case; it was a downright provocation. Ever since the Norwegian parliament agreed to make the "environmental" paragraph 112 part of the constitution, its moral core has been viewed as a recognition in law of the principle that no generation has the right to degrade nature to an extent that harms the life prospects of future generations. Given current knowledge about what sort of practices contain a high probability of causing harm to our successors – namely the extraction and use of fossil fuels – the duty on present generations(s) is sharpened: we cannot pretend not having known about the sort of harm we cause if we neglect to implement the changes (i.e., reduction of emissions) required. From a moral point of view, then, it is unacceptable if we who live today, knowingly and with eyes wide open, choose to go on overexploiting the natural world so that future citizens are denied their equal right to a nature that is flourishing and rich in biodiversity. The issue boils down to a principle that a variety of world views and ethical theories agree on: it is unacceptable that some people exercise their freedom in a way that undermines the basic needs – life prospects – as well as liberties of others.

Thus understood, the moral principle that forms the core of paragraph 112 appears to have been violated in the Supreme Court's verdict, unequivocally and shockingly so. How could this happen?

Let me first say that a case where a verdict contradicts a morally compelling principle such as the present one is deeply problematic. It raises questions about whether the duties and the rights stipulated in the law – duties of present humans, rights of future ones – are devoid of legal merit and significance. To be sure, the issue of the proper relationship between morality on the one hand and positive law on the other is too comprehensive to examine here (see Habermas 1996; Lysaker 2024). More specifically, the verdict we are discussing demonstrates how the court, with seventeen judges – and the judgment was nearly unanimous – in effect pronounced on their view on how they assess the relationship not between morality and the law, but between the law and politics. The argument is that were they to accept the claim made by the environmentalists and so demand that continued oil production be stopped, the court would "interfere" with policy decisions that properly belong to the political authorities, based on the interplay between parliament and government. In other words, the court needs to "keep an arm's length" from politics. Thus, the argument goes, the tripartite division of power within a liberal democracy is upheld and respected.

Whatever the merits of this argument in purely principal terms, I suspect that the reasons for the court decision to acquit the Norwegian state lie elsewhere. Consider what is at stake: a verdict in favour of the environmentalists' claim would mean that, by implication, various Norwegian governments since 2014 have been responsible for implementing oil policies that are in breach of the constitution. Nothing less. Viewed in this way – and it is a realistic way of looking at it – a guilty verdict appears implausible, if not virtually impossible – for *political* reasons at that. Why, exactly, is that so?

The answer is not in terms of the conflict between a minority and a majority as conceived by Rawls, explained above. If anything, a *greater* number of people (to stick to Rawls's anthropocentrism) are likely to be harmed by present-day and near-future emissions than those living today and so able to articulate their opinion on the matter. Indeed, temporal as well as spatial dimensions of the situation is such that *evermore* people in *evermore* decades to come and in *evermore* places on the planet are adversely affected by current practices of extraction. A minority of people – the most affluent 10 per cent of today's world population are responsible for more than 50 per cent of emissions – are causing this *cumulative* harm, quantitatively as well as qualitatively speaking; more people will be harmed in a future 2, 5, or 3, 0 degrees warmer world than in today's 1, 3 plus warmer one, the latter world being more harmful than the former one. So again, the assumption that is not unique to Rawls but representative of mainstream approaches to civil disobedience does not hold up here: whereas those who actually take part in such actions may well be "a minority" in present circumstances, in the case of protests over climate policies the party on whose behalf this minority takes to the streets is *de facto* the overwhelming majority of human persons. Climate rights are not minority rights; climate justice is not justice for some minority. To be sure, the Rawlsian liberal argument about a minority's right to protection is a perfectly sound principle on its own merits. Only it is not the principle at issue here.

In suggesting that the verdict has to do with politics rather than principles, a comparison with a similar case is instructive. In the Netherlands, environmental organizations sued the state for violating the obligations under the Paris accord, signed by the government. They won. In the so-called Urgenda case, in a verdict catching world-wide attention in 2019, the Dutch Supreme Court decided that the plans for emission reductions put forward by the government fell short of the targets signed up to in Paris; the policies had to be revised accordingly (see Krznaric 2020: 186). Similar cases have been successful in countries like Ireland, France, and Germany, all well-organized liberal democracies. So why does Norway differ?

The reason is obvious: revenues from the production of oil in Norway are a main contributor to state finances, and the oil and gas lobby, led by Equinor, is by far the most powerful one in the country, tirelessly repeating the assertion that the Norwegian welfare state depends on the income brought by oil and gas, besides employing approximately 200 000 people (subcontractors included). Norwegians born after 1970, the year Equinor (named Statoil at the time) started drilling in the North Sea, have grown up with this narrative according to which, were it not for its role as a major producer and exporter of fossil fuels, Norway would never have been able to finance the social democratic welfare state for which the country is admired world-wide, and of which Norwegians across the generations are so proud. This narrative is a gross distortion of history. The fact that is suppressed to such a degree that the younger generation is completely ignorant about it is that this welfare state was put into place, by political agreement and cooperation across the aisles, during the 1950s and 1960s, that is to say, in the decades *prior* to the discovery and commercial extraction of oil within Norwegian territory. Important as it is to set this record straight, not least for Norwegians themselves (!), the policies of various governments since the 1970s have seen to it that the narrative has become self-fulfilling: the state's reliance on oil and gas, on increasing their volume and on these fossil fuels being in persistent demand and catching good prices on the world market, has lent a growing proportion of truth to the initially misleading claim about the welfare state's petro-dependency. For lack of a future-oriented diversification of its economy and the sort of jobs offered to a new generation, by laying too many eggs in the oil basket in particular, the Norwegian state has de facto become ever more vulnerable to fluctuations in fossil fuels markets, in effect – if not by intention – staking its continued prosperity as a twenty-first-century welfare state on oil and gas extraction *continuing to grow rather than abating*, in blatant contradiction with the Paris Agreement.

In Chapter 2, when discussing Kari Marie Norgaard's work, I mentioned that young Norwegian climate activists are often portrayed as "traitors", as betraying the legacy of previous generations' efforts to provide a high standard of living young people today simply take for granted, revealing themselves as spoilt and ungrateful. The court case discussed above illustrates that this critical, if not downright hostile, attitude toward climate activists not only pervades public perception of their acts of civil disobedience but is manifest in court decisions as well. Tellingly, the same acts of disobedience are punished two or three times more harshly in Norway than in Sweden; this holds for the fines imposed for actions such as obstruction of traffic, making for a conspicuous contrast between two neighbour countries whose regimes of law and punishment are very similar. So again, why does Norway differ? The short answer is that Sweden is not an oil-exporting country.

4.3 Malm's historical lessons

Given that the fossil industry in a country like Norway is so successful in blocking endeavours to force the authorities to reduce its contribution to climate change, activists need to reconsider their options. They need, that is, to make decisions about what types of protests to engage in to show their readiness to break the law, making manifest their conviction that what is regarded as acceptable and legal today, will be judged as unacceptable and illegitimate tomorrow. Following this conviction, they look upon themselves and the role they volunteer for as that of whistleblowers, directing the public's attention to dangers that threaten the community, demanding that urgent action be taken, in keeping with the principle of precaution.

When politics and business as usual are seen as jeopardizing the future rather than securing it, how to go about protesting? Is there not a valid concern that activists will be radicalized? That at least some of them will give up on peaceful methods that, for all the modesty and discipline they exhibit, either simply are ignored by politicians and the public alike or end up with a negative outcome in the courtroom? When nothing less than the future of the planet is at stake, what to do? If ordinary methods have been exhausted and tried in vain, why not try extraordinary ones?

In his book *How to Blow Up a Pipeline*, the Swedish ecology scholar and activist Andreas Malm argues in favour of certain types of sabotage against property and infrastructure crucial to fossil energy operations. He makes a sharp distinction between actions aiming to damage material entities (i.e., oil- and gas-installations, pipelines), and actions meant to harm – or involving a calculated risk to harm – human beings and animals. The distinction between property and living beings is not merely descriptive, of course; it is eminently moral, and carries political implications.

Malm is careful to point out that he does not endorse violence or any kind of damage that targets humans and animals. Even so, the title chosen for his book is deliberately provocative, and some readers might be willing to go further than Malm himself with regard to the types of actions viewed as both necessary and legitimate.

Offering a brief history of civil disobedience, Malm distinguishes between two types of pacifism. The first is *moral* pacifism, holding that it is always wrong to commit acts of violence. He quotes Bill McKibben, prominent climate activist in the United States and founder of 350.org, who promotes the philosophy of "turning the other cheek", citing Martin Luther King Jr. as a model. On this view, whose theological origins are evident, there is a "spiritual insight" at the core of non-violence, affirming the importance of "taking on unearned suffering" and seeing it as "redemptive". Malm is critical of this view, asking "why would it be noble to subject oneself to suffering one doesn't deserve? Isn't suffering unearned by the victims precisely what is so morally repugnant about the unfolding [climate] crisis? If so, why make it a virtue?" (2021. 33, 34).

The problems Malm sees with moral pacifism bring him to consider the second type, *strategic* pacifism. Rather than giving a principled argument for pacifism, it is recommended in an instrumental manner. Non-violence, the argument goes, is superior to violence by being a more efficient means to reach the goals aimed for; as a method it is less counter-productive, less dangerous, and less contentious. Roger Hallam, one of the founders of Extinction Rebellion in England, puts it like this: "Violence does not optimize the chance of successful, progressive outcomes. In fact, it almost always leads to fascism and authoritarianism. [...] Non-violent discipline is rule number one for all participants" (Malm 2021: 34f.; Hallam 2019: 101).

Historically, however, the picture is not at all as unequivocal as the statements made by McKibben and Hallam suggest. Malm aptly recalls the protest actions carried out by the British suffragettes led by Emmeline Pankhurst, under the slogan "Deeds not words": "The argument of the broken pane" involved a large number of well-dressed women smashing every window they passed on the street. In other words, the destruction of property, carefully selected such, public no less than private, was a crucial element in the actions seeking to achieve the political goal of the women's right to vote. "To be militant in some form, or other, is a moral obligation", Pankhurst told her co-protestors. Shifting attention to the civil rights movement in the United States, destruction of property was once again a frequently used means to advance political demands. For example, in 1967, more than thousand businesses were damaged or destroyed in Newark alone. When Martin Luther King Jr. was arrested following the demonstrations in Birmingham in 1963, he sent the following message from his prison cell to the Kennedy administration: if the demands of his movement were not met, "other, more menacing forces would arise", so that, if the channel of non-violence remained closed, millions of Negroes, out of frustration and despair, would take to the streets, causing them to be "flowing with floods of blood" (2021: 48f.). Only when President John F. Kennedy understood that public order was threatened, was he prepared to take their demands seriously.

The lesson Malm takes this example to teach us is that the emergence of a "radical flank" within the protest movement may serve a vital purpose. By referring to the likelihood that "less responsible leaders" – notably Malcolm X – would gain momentum if the reasonable demands made by the "moderate" majority of the protestors were to be denied, King skilfully exploited the President's worry that a small group of activists would turn to violent methods threatening the public order, and so undermining the President's reputation. Without saying so, King invited Kennedy to be partners in a "responsible" alliance, having a common interest in minimizing the danger of escalation into violence. Malm puts the lesson like this: "The civil rights movement won the Civil Rights Act of 1964 because it had a radical flank that made it appear as a lesser evil in the eyes of state power" (2021: 49). In other words, "without Malcolm X, there might not have been a Martin Luther King Jr. (and vice versa)" (50). Malm makes a similar case for the tactics used by the African National Congress (ANC). Giving the lie to the selective memory according to which Nelson Mandela was a principled pacifist who never promoted or practice violent methods, Malm cites the would-be Nobel Peace Prize laureate's statement that "our policy to achieve a non-racial state by non-violence has achieved nothing", elaborating that "if sabotage did not produce the results we wanted, we were prepared to move on to the next stage: guerilla warfare and terrorism" (2021: 52).

For all the fascination they hold, how relevant are these historical examples to present-day climate activists? Must we not admit, Malm asks, that the struggle against fossil fuels is of a wholly different character than, say, the struggle against slavery, against male monopoly on the vote, and against apartheid? Indeed, are not the differences so big that the present struggle "will succeed only on condition of utter peacefulness?" (53).

Before answering this question, Malm invites us to consider that the climate emergency is not only different from the struggles mentioned; it is worse. What we need to recognize, then, is how profoundly climate change is anchored in the very structure of contemporary society: politically and economically, to be sure, but also, and no less importantly, mentally and morally. These are my terms, not Malm's, but they are mirrored in what he does say: "Fossil fuels are not a political arrangement like limited franchise or pass laws: they and the technologies they power are *productive forces* imbricated in certain property relations" (55). Making his point in unmistakably Marxist terminology, Malm maintains that "business-as-usual is not a sideshow to bourgeois democracy"; "it is the material form of contemporary capitalism, neither more nor less" (56).

Unfortunately, in another instance of bad timing, those who engage in the struggle against fossil fuels, understood as a struggle both *against* and – inevitably – *within* the deep structures of contemporary capitalist society, have to agree on its methods in a situation shaped by what Malm calls "the demise of revolutionary politics". Such politics, he points out, "barely exists any longer as a living praxis in powerful movements or as a foil against which their demands can be set" (61). Malm's grim diagnosis echoes that given by Mark Fisher and Dany-Robert Dufour, discussed in the previous chapter. Often much too sweeping, the task for Malm is to make that diagnosis – the death of radical politics – instructive to the current crisis, different from previous ones. "With the consequent deskilling of movements", observes Malm, "comes the reluctance to recognize revolutionary violence as an integral component. This is the impasse in which the climate movement finds itself: the historical victory of capital and the ruination of the planet are one and the same thing" (62). As if that were not enough, there is another feature to be cognizant about, one that more than any other sets the climate crisis apart from others: time. The level of urgency is different from that involved in what may be called "traditional" conflicts and struggles as conceived in mainstream political science and in Marxism alike: they have typically been fought between human actors, or classes, over the just distribution of rights and life prospects. Distinct from this, the urgency of the struggle against fossil fuels involves us humans in a crisis not only against – or with – *each other*, but where *the natural world is itself the locus of crisis*, not the only locus, to be sure, but one that impacts on all the conventional loci and causes of conflict.

Being by his own admission a Marxist, the question is whether Malm is better equipped – philosophically and politically – to recognize the role *nature* plays in the present crisis than is the tradition of Marxism's major rival, political liberalism as conceived from Locke to Rawls. This is a big issue in its own right, largely implicit in Malm's *How to Blow Up a Pipeline*. I shall say more about it below.

When discussing what methods and tactics are appropriate in today's struggle against fossil fuels, Malm's Marxist credentials come to the fore in his insistence on the pivotal importance of private *property*. "Capitalist property", he contends, "has the status of the ultimate sacred realm" (2021: 68). He quotes, approvingly, the scholar R. H. Lossin's statement that "sabotage is a sort of prefigurative, if temporary, seizure of property". With regard to the climate emergency, sabotage is "bothy a logical, justifiable and effective form of resistance and a direct affront to the sanctity of capitalist ownership" (68). Property, Malm points out,

> does not stand above the earth; there is no technical or natural or divine law that makes it inviolable in this emergency. If states cannot on their own initiative open up the fences, others will have to do it for them. Or property will cost us the earth.
>
> (68f.)

To be sure, the latter is a hard-hitting slogan, but it doesn't answer the question about what to do, given the severity of the crisis. Malm reminds us that 80 per cent of the energy used in today's world is fossil, just like it was when the first COP took place in Rio De Janeiro in 1992 and when the UN panel on climate change was established. How tall an order change truly is, is borne out in the fact that the global energy system is the largest network of infrastructure ever built. This being so, sabotage from a militant wing of the climate movement will not bring about the transition required, on the scale needed, and when the window is rapidly closing. The states must come forward and assume responsibility – yet they are doing no such thing. For Malm, the states – and this is the general way he speaks about them – are clearly a part of the problem rather than the solution. He therefore proceeds to concentrate on what activists can do when state actors and governments fail to engage on the correct side of the struggle, remarking that "protest is when I say I don't like this. Resistance is when I put an end to what I don't like" (70).

Malm draws attention to several instances where extensive property destruction has been carried out, successfully so. The Movement for the Emancipation of the Niger Delta (MEND) is a case in point. In "Operation Cyclone" between 2006 and 2008, the group blew up pipelines, stroke vessels, overpowered offshore platforms, assaulted officers, and kidnapped oil employees, forcing a third of production in Africa's largest oil country Nigeria to shut down. Shell and ExxonMobil were "on the verge of withdrawal", comments Malm (2021: 74). He mentions similar acts of sabotage in Egypt, India, Yemen, South Africa, and Colombia. Indeed, in recent years they have occurred closer to Malm's home, in Scandinavia and in England and Germany. Malm ends his book by taking the step from scholarly chronicling the story of such activism to sharing his experience as a participant, having been involved, alongside thousands of others, *in situ* in the "Ende Gelände" that targeted a big coal mine in the eastern region of Lusatia in Germany in 2016. Vattenfall, the company operating the mine, had to declare that the sabotage had enforced the suspension of all electricity production, "something that had never happened before at a fossil-fuelled power plant in Europe" (159f.).

Given this record of well-organized acts of sabotage that have proved successful, the question is: is this a recipe for how to bring about the lasting structural transformation needed? To be sure, and as we witness, it is technically possible for people "organized outside of the state to destroy the kind of property that destroys the planet" (75). And yet, again and again, for all the spectacular success and media attention that go with some of these acts of sabotage, the shut downs they enforce are temporary. As a rule, once the activists and the media leave the scene, operations resume and business is back to its catastrophic normal.

Small wonder that this is so, and most likely will remain so in the foreseeable future. Why? Consider the massive world-wide Goliath that so many Davids dispersed around the continents are up against, namely a global energy system that is "the largest network of infrastructure ever built", to recall Malm's own words. As far as the status enjoyed by private property is concerned, the hegemony of neoliberal ideology helps see to it that such property is viewed as unproblematic, indeed a source of progress and prosperity by a majority of people in the most affluent countries in the world. Of course, these are the very countries that continue to contribute most to emissions, and so to *resist* demands – be they of the politics as usual kind, or be they advanced by activists of various strands – to engage in the transition needed. Indeed, the *resistance* to the transition appears to be much more well-organized and well-funded as well as culturally entrenched, for reasons I have explored in previous chapters.

How, then, does Malm tackle the fact that the resistance to change the system is so much more powerful, along mental as well as material dimensions, than the acts of resistance hoping to bring it down?

To be sure, to put it like this is to go back to start: Malm's point of departure is the "reality check" that tells him that the system is in no way about to change in the way required to "solve" – as the rhetoric of sustainability has it – the climate crisis. So how to convince a majority of people to shift from supporting the system to supporting, or even joining, the presently marginal actors who devote themselves to the struggle? To his credit, Malm responds to this question in a more concrete way than theoretically minded Marxist scholars tend to do. He realizes that for all the "principal" importance of the institution of private property, attacking the problem at its root causes must involve the role played by private *consumption* as well as the *production* of fossil fuels; it must, that is, embrace both the favoured focus of liberal discourse and that of leftist one. Taking the targeting of Volvo SUVs in Sweden as his example, and again speaking from first-hand experience, Malm is quick to assure us that his Marxist stance is once again useful, pointing out – by way of expensive cars – that "the richest 0. 54 per cent of the species emit one-third more than the poorest half" (2021: 86). So yes, consumption is no less important than production and the issue of who owns the infrastructure and makes the profits from the persistent dominance of fossil fuels; only it is the consumption of the richest one per cent, or ten per cent, that by far counts the most. In this regard, the distinction between luxury and subsistence emissions is crucial, catching the grotesque disproportionality of culpability among present-day humans who "all" contribute to the climate crisis. Rarely, if ever, has a crisis upheld by a society's deep structures, politically and economically been pervaded by such a degree of inequality among both those helping cause the crisis and those affected by it.

To say this is to invite a comparison with slavery. The popularity of this comparison is not only amply documented in the literature; it is also attested to by the frequency with which it is being brought up by all sorts of audiences when I give talks about climate change. As for Malm, he discusses the struggle to end slavery as a mandatory part of the canon of resistance. Slavery, he notes,

> was not abolished by conscientious white people eagerly disassembling the institution. The impulse to subvert it sprang, of course, from the enslaved Africans themselves, and they very rarely possessed the option of non-violent civil disobedience; staging a sit-in on the field or boycotting the food offered by the master could only hasten their death.
>
> (2021: 39)

The Haitian Revolution was the first emancipation of slaves; in the United States, slavery was terminated by a bloody civil war; apartheid in South-Africa was ended after years of resistance involving methods far from peaceful in the sense of the – highly selective – way in which Nelson Mandela is remembered. In all these cases, "collective action against slavery perforce took on the character of violent resistance" (39).

The lesson from history, then, is that slavery ended thanks to the resolve of the slaves and their allies to fight back, employing a diversity of methods, some of them violent. On Malm's view, the readiness to use violence has proved decisive in securing success. But what about morality? Is it possible to mobilize the activism required, both among those directly affected and among various outsiders, without a compelling argument that slavery is *morally wrong*? Wrong, that is, to such a degree that it doesn't help that slavery enjoys the status of a legal institution? That, indeed, its de facto legality, regardless of how long it has lasted, only makes it worse, necessitating that *what has been legal be regarded as illegitimate*, insofar as slavery is looked upon as morally repugnant? That, if we do not put an end to it, we will be judged as both morally wrong and on the wrong side of history?

To be sure, we have been here before. I mentioned my argument as an expert witness in the case involving climate activists – the argument, that is, that for practices (such as expanding fossil fuel installations) to be legitimate, it is not sufficient to point out that they are de facto legal. Rather, true legitimacy can be lent to such practices not by lawfulness alone but, more basically, by their being morally right, only then *deserving* our allegiance to them. Practices, and the institutions upholding them qua legal, need to meet the requirement of being just and so desirable for all affected parties (to allude to Habermas's way of making the argument; 1990, 1996). Malm does not engage in a philosophical discussion about the foundation of legitimacy in morality. He restricts himself to a reference to Robin Blackburn, a well-known Marxist scholar of slavery, to the effect that

> the very notion of slavery as unethical – harmful to the slaves, whom the masters wished to portray as happy and docile – originated in the acts of explosive refusals. Even the most pacifist Quakers pointed to the revolts as proof of the horrors of the peculiar institution.
>
> (Malm 2021: 40)

Since both Malm and public discourse make the comparison between slavery and the climate crisis, it is vital to be precise about what we are comparing. Is there a valid analogy between coming to recognize that "for one human being to own another is morally wrong" (Callicott 1999: 33) and coming to recognize that the practices allowing the continued extraction of fossil fuels are morally wrong? Whereas the first wrong takes place between human beings, the second is a case of a wrong done not only by some humans against other humans, but against nonhuman creatures as well – consider that during the last 50 years, 73 per cent of wild animals (individuals, not species) have disappeared due to human activities. So, if a morally grounded indignation drove the effort to end slavery, deeming it a gross injustice among humans, what would spur the effort to end the epoch of fossil fuels? What is required to mobilize the support needed for this struggle to succeed?

To ask such questions is to realize that, for all the recalcitrance of the institution of slavery and the bloodshed it took to bring it down and declare it illegal (not forgetting that even today slavery still exists in many parts of the world), the analogy with the current environmental crisis runs into a number of difficulties, rendering the differences more pertinent that the similarities. The all-important point is this: Whereas the "moral revolution" at work in fighting slavery was internal in the sense of being a shift (revision) *within* and so loyal to the anthropocentric framework, insisting that equal dignity and worth apply to *all* human beings, not only some, to the exclusion of others, current fossil fuel practices are not only an instance of injustice among humans – with some profiting, others suffering – but also a matter of doing wrong – irreversible harm – to *nonhuman* nature and all the creatures and lifeforms involved; a wrong that in its consequences amplifies the sufferings of an increasing number of people around the world, since the degradation of the natural world degrades the conditions of life for human societies dependent on it.

So what kind of shift is needed in the case of climate crisis? Does it suffice to remind the powers that be, whose power often relies on fossil fuels, that to continue down the current path will result in growing human misery, forcing millions or billions of people to move because their livelihood is endangered? That is to say, does it suffice to reiterate the "internal" anthropocentric argument that was made with such (albeit incomplete) success in the struggle against slavery, invoking the morally unacceptable harms inflicted on human beings affected by climate change? Or do we have to realize that the anthropocentric framework, admitting dignity and intrinsic value to humans and humans only, is part of the problem – so much so that it needs to be replaced by a moral framework including all kinds of living beings (biocentrism), or the entire natural world, encompassing inorganic entities as well as organic ones (ecocentrism) (Lysaker 2024)? In other words, is the environmental crisis of such a nature that our ambition, in terms of an overall shift in values and attitudes, must be nothing less than that "the wanton destruction of the nonhuman world will come to be regarded as equally unconscionable as slavery", to quote the environmental philosopher J. Baird Callicott (1999: 33)? Or is this too tall an order? Unrealistically tall? Or unnecessarily tall?

In my previous books (2015, 2019, 2023), I have argued that as a society we need to replace anthropocentrism with a framework that includes, at least, all kinds of living beings, holding that practices of ours that degrade nature constitute a wrong done *to* nature and not only to the humans harmed. My major thesis is that the conditions of life define the conditions of ethics and politics, so that making it possible for life to go on for all sorts of living beings defines the task of ethics, its *raison d'etre*. In short, what matters ethically (normatively) cannot be separated from what matters ecologically and biologically (factually); yet this separation, that of fact from value, and that of the individual from the whole, is what theories based on methodological individualism, implying ontological such and resulting in ethical such (individualism) are committed to – mistakenly so. The philosophical argument for my position is not what concerns me at this juncture, however (see Rolston 1988; Vetlesen 2015, 2023b). What we are presently discussing, in the light of Malm's position on resistance, is the political question of how to mobilize support for the morally just cause of bringing the fossil fuels epoch to an end, the sooner the better.

On my reading, Malm does not regard it as necessary to go beyond anthropocentrism: for him, as a Marxist, capitalism, with its systemic twin imperatives profits and growth, is the problem, the root cause of the climate crisis, not anthropocentrism. The fact that capitalism as we know it, since its beginnings some four hundred years ago, has taken anthropocentrism's idea that nonhuman nature is mere means to human ends for granted and thus helped maintain its hegemony in practice (albeit in the exclusivist manner compatible with slavery) is not something that Malm pays attention to. His Marxist credentials also shine through when he accords primacy to what David Harvey (2010, 2014), the Marxist geographer, calls the first contradiction of capitalism, that is, the conflict residing in the *capital–labour* relationship, this being seen as the conflict ripe with pathologies of alienation and of crisis, hence with the potential for labour's resisting and toppling the system. The second contradiction, by contrast, resides in the *capital–nature* relationship, whereby the potential for crisis lies in capital's compulsive expansion increasingly coming up against limits, or barriers, or scarcities in nature, such as cannot be avoided or circumvented, at least not in the long run. Since capitalism cannot stop expanding and still be capitalism, its trajectory is self-destructive: Today, capitalism's word-wide exploitation of nature has become so wholesale as to spoil and destroy that which is being exploited at its core, in its capacity to reproduce itself and to replenish; *overshoot is systemic*, not accidental, temporary or local. This holds for both types of exploitation: the burnout in/of human capital (labour) as well as the overshoot in/of natural capital. The point, much more evident today than when Marx wrote *Capital*, is that capital's exploitation of labour crucially and non-optionally depends on labour's exploitation of nature; the former – always the key focus in Marxism's notion of conflict and crisis – cannot persist, hence profits not be made, growth not continuing, without the latter, the intact substratum of nature, man's and capital's "other". This means that the Marxist tradition gets it backwards when it grants the capital-labour constellation primacy over nature; the constellation is in fact parasitic on nature, and can only persist on the condition that nature is not exploited beyond repair.

To be sure, to speak about "the Marxist tradition" as were it one homogeneous body of thought, with a single unified stance on these issues, is unacceptably simplistic. The time when Marx's oeuvre was regarded, by advocates and critics alike, as it were, as subscribing to the "Promethean" position committed to man's domination of nature, to technological optimism (if not determinism), to productivism and the ideology of progress – so that the more effectively nature is exploited for human purposes, the greater the prospects for humanity, provided the benefits be fairly distributed among the classes – is gone. In recent years, scholars such as Fred Magdoff, John Bellamy Foster (2000, 2011), James O'Connor (1998), and Kohei Saito (2022) have made the case for an "ecological Marxism", pointing in particular to the passages where Marx addresses the "metabolic rift" between humanity and nature, exemplarily tied to the overexploitation of soil and as the rift in nutrient flows between urban and rural areas.

Was Marx, then, a Promethean, or was he one of its pioneering critics; or a bit of both, perhaps contradicting himself or not quite making up his mind? This is not the place to try and settle the question; it would take a separate chapter, if not a book, to do so. That said, I consider Joel Kovel's assessment in his study *The Enemy of Nature* a well-informed and nuanced one, in many respects sympathetic to Marx, yet pointing out that "yes, humanity is part of nature for Marx; but it is the active part, the part which makes things happen, while nature becomes what is affected by human activity" (2007: 232). Except for some scattered passages in various writings, continues Kovel, "nature to Marx appears directly as use-value, as not as what use-value leaves behind, namely, recognition of nature in and of itself. In Marx, nature is subjected to labour from the start"; his conception of labour "involves an essentially active relationship to a kind of natural substratum" (2007: 232). As for the political regimes which have called themselves "Marxist", or Marxist-Leninist or -Maoist, Kovel remarks that "almost the entire socialist tradition, including those branches of it unburdened by Stalinism, has largely been unable to appropriate an eco-centric attitude" (2007: 226). I agree with Kovel's assessment, and I have criticized Marx's reservation of the source of value, of its locus, to human activity – exemplarily, labour – with its implication that nature as "untouched" by us and our interests is itself mute, devoid of intrinsic value (see Vetlesen 2015: 32-40; Brennan 1993, 2003; Vetlesen 2016). Alf Hornborg, Malm's mentor and colleague, sums it up well: "A coherent ecological Marxism must acknowledge that other productive inputs than labour can be sources of capitalist surplus", inputs comprising "a wider category of biophysical resources including embodied energy, labour, land and materials" (Hornborg 2019: 173).

To return to Malm, his version of Marxism is not one that commits him to go beyond anthropocentrism so as to critique the destruction capitalism inflicts on nonhuman nature as a moral wrong *to* that nature, since violating the loci of value *integral* to nature; he restricts that category to harms done to humans. In his breakthrough as a scholar, the historical masterpiece *Fossil*

Capital: The Rise of Steam Power and the Roots of Global Warming, Malm addresses the issue of resistance in the last chapter. "The arithmetic of climate change", he writes, referring to the scary fact that emissions grow faster and faster when they desperately need to be reduced, "tightens the screw on Marxists as much as on everyone else. Any argument along the lines of "one solution – revolution" or "socialist property relations are necessary to combat climate change" is now untenable" (2016: 383). So what follows? "At this moment in time", Malm continues, "the purpose of an inquiry into the climatic destructivity of capitalist property relations can only be *a realistic assessment of the obstacles to the transition*. They grow higher by the day" (383). He concurs with one of many reports that conclude that "it seems unlikely that governments will undertake emergency mitigation, even when life-threatening climate disasters occur" (385). Commenting on the rise of geoengineering, defined as "deliberate, large-scale interventions in the climate system designed to counter global warming or offset some of its effects" (386), Malm laconically remarks that "planning the economy is the ultimate taboo; planning the climate is worthy of consideration". Indeed, "it has become easier to imagine deliberate, large-scale intervention in the climate system than in capitalism" (388).

In summing up the situation like this, Malm is true to his promise to be realistic about the obstacles to the transition called for. Echoing our earlier discussion, he acknowledges that resistance to change is bolstered by how consumerism encourages us – people of all walks of life, across divisions of class – to express our identity through our material possessions, whereby losing these may mean losing our sense of identity and self-respect. Yet efforts made to persuade consumers to change their habits, to buy "green", to use public transport, and so on, are mistaken; they deflect attention, contends Malm, "away from production, the active moment determining business-as-usual as a whole *including* the widening circle of consumption". Hence, as a general compass, "the targeting of the consumer leads into the classical blind alley of Western environmentalism" (365); it displaces culpability from the system to the individual, in tune with neoliberalism's favourite formula for problem-solving in general, and recalling the analysis in Chapter 3 of how the imperative to build resilience bolsters the primacy of personal responsibility over structural change.

Malm takes issue with Dipesh Chakrabarty's influential claim that "the climate crisis is not *inherently* a result of economic inequalities", since "the lurch into the Anthropocene has also been globally the story of some long anticipated social justice, at least in the sphere of consumption. This justice among humans, however, comes at a price" (Chakrabarty quoted in Malm 2016: 390f.). What price, and for whom? For humanity, according to Chakrabarty, in that "humanity is unified not only as the source but also as the victim of this crisis"; the human species is "a universal that arises from a shared sense of catastrophe" (391). One does not have to be a Marxist to reject this universalism – all humans are in the same boat – as false. Writes Malm:

> Witness Katrina in black and in white neighbourhoods of New Orleans, Sandy in Haiti and in Manhattan, sea level rise in Bangladesh and in the Netherlands, all the realities of differentiated vulnerability in any impact of climate change, direct or indirect.
>
> (391)

If anything, the harms inflicted by way of climate change pay testimony to the ever-greater difference made by class-related inequalities both between and within the communities affected.

4.4 Criminalization and anger

Given this class-oriented analysis of the causes of the crisis, what does Malm recommend for resistance to be effective? He reiterates his assertion that the right to property must be broken, only to add that "the order-keeping state stands in the way". The fact is that the rich drive the emergency, "and a climate movement that does not want to eat the rich, with all the hunger of those who struggle to put food on the table, will never hit home". Indeed, "not only do the rich make our lives miserable, they are working to *terminate* the lives of multitudes" (2021: 127f.; italics in original). What is badly needed, then, is a resistance movement motivated by the amount of "social anger" that the disregard of the rich for other people calls for. But such anger is not forthcoming; instead the climate movement is "placid and composed" (129).

I discussed resistance against resistance above, and a frequent component is anger – not among the protestors, but among those whose everyday life is disrupted by their actions. XR activists disrupting rush-hour traffic in cities like London or Oslo are a case in point. Such protests typically trigger anger among those affected, leading to a backlash. Yet a discussion of these lessons is missing in Malm's book. He notes the growing trend to criminalize non-violent climate protests, this of course being the response of the state and not of ordinary people, whereby it must be said that polls show a majority of the public to support more draconian sanctions against activism. Increasingly, sabotage in the sense approved by Malm – the damage and destruction of property, like oil pipelines, where no blood is shed – are labelled "terrorism", and punished accordingly. Marking non-violent protests as "non-democratic" is very much part of this tendency; rather than being welcomed as vital participants in a democratic order open to criticism, say, in the liberal style of Rawls set out above, activists are treated as state enemies (Trump: "internal enemies"), not fellow citizens whose cause merits a *political* response, as opposed to a punitive one. Indeed, the criminalization of non-violent protest through exception (to allude to Agamben) diverts attention from underlying social conflicts – over energy sources, over land – and makes for a regime of securitization such that politically motivated organized protest *as such* is delegitimized and depoliticized (see Weis 2022: 23f.). In this way, and through increasingly harsh "pre-emptive" police methods including infiltration and around the clock surveillance, young activists – often students – are painted as "suspect", as "not one of us", as "the enemy within" against which the state as well as "ordinary, law-abiding citizens" have the right to be protected. Such active "othering" hampers the activists' efforts to appeal to the public at large, to reach out for support across divisions of class and to bridge growing differences between rural and urban areas.

In recent years, I have witnessed how various elements that characterize criminalization of peaceful climate protests have come into force in my native Norway, taking the harsh judicial regimes of countries like the United States, the United Kingdom, Germany, and Italy – and not blatant authoritarian ones like Russia or China – as the model to copy, rendering the difference between "liberal" and "authoritarian" nations increasingly inconsequential, merely rhetorical, as far as the treatment of protestors is concerned. In a report submitted by Michel Forst, the UN's special rapporteur on environmental defenders in January 2024, he notes that "draconian anti-protest laws", massive sentences and court rulings forbidding protestors from explaining their motives to juries are "crushing fundamental freedoms". By way of example, Forst points out that "until recently it was very rare for members of the public to be imprisoned for peaceful protest in the UK". As George Monbiot comments in *The Guardian*, "now you can get six months merely for marching". The message sent is clear: "The powerful are protected while the powerless are exposed to ever more intensive laws. Economic inequality is mirrored by inequality before the law" (Monbiot 2024: 46).

How to respond? We recall that Malm thinks highly of the so-called "radical flank effect", viewing confrontational charismatic activists like Malcolm X as instrumental in securing that the comparatively "moderate" Dr. King gets a hearing from the President, because otherwise – the reverend will warn – protesters will get really angry, and so willing to turn militant and, probably, more violent. Yet Malm realizes that "extremism can make a movement look so distasteful as to deny it all influence" (2021: 121). Today more than ever, what is to be labelled – branded, stigmatized – as "extremism" is a matter of perception; and the powers that be tend to be very good at shaping that perception, as the "othering" of young activists in particular shows.

As with the fight against slavery, climate activists – especially the young – strongly believe that they have morality as well as history on their side.

Claiming the high moral ground on behalf of their cause, insisting on their right to inherit an ecologically inhabitable planet, their response to the draconian anti-protest laws presently implemented in one "liberal" democracy after the other is that the "organized crime" governments need to deal with is that committed by the fossil industry, pushed by tireless well-financed corporate lobbying. Again, what is considered perfectly legal, and what brings the greatest profits, is the speeding up of the fossil-based production and consumption patterns that will lead to runaway climate change, whereas what is treated as illegal and punished as such are attempts to avert this catastrophe. To put the "moral of the story" as it now unfolds in such terms, is to invoke the legitimacy of anger, indeed the need for it – for anger to be admitted, to be felt, and to be displayed in public, so as to call for a response. For lack of going public, being shared with others, being directed at somebody or something in particular, anger will stay with the individual in question, being privatized, a psychological issue, when it urgently needs to be politicized.

Recall Malm's statement that "a climate movement without social anger will not acquire the required striking capacity, and it should have no difficulties developing the point", citing the slogan "More ice sheets, fewer bankers" (2021: 127f.). He then adds the sentence I quoted above: "Not only do the rich make our lives miserable, they are working to *terminate* the lives of multitudes" (128; italics in original). Here lies the task for the radical flanks of the movement, asserts Malm: "those who dare to speak the name of the enemy" (128).

This may appear compelling on paper. But is it working? When attempted in practice, how is the response from the wider public? Is the anger Malm wants to see go public being picked up? Does it mobilize people, helping the movement grow?

Of course, there is not and cannot be a simple answer, valid in all cases. If we stick to the experiences made in countries like the United States, the United Kingdom, and Norway, acknowledging the differences in political cultures, my impression is that many people do get angry – only they direct that anger at the protestors, not the bankers. Indeed, the rise of right-wing populism in recent years (the Tea Party movement and Trump's MAGA in the United States, the Reform party in the United Kingdom, etc.) is inseparable from the mobilization and the giving vent to accumulated, individually felt and socially shared anger (see Hochschild 2024, discussed in the Introduction), directed at a selection of culprits perceived as responsible for a series of setbacks suffered, as well as at individual activists – Greta Thunberg is a particularly illustrative case in point; more on which below. How to explain this anger directed at protestors? Why do their activism meet with strong negative feelings instead of sympathy and support? Shouldn't we expect adults to identify with the anxiety and to acknowledge the desperation that many young people – be it their children, or grandchildren – articulate in the public sphere? What does the lack of support express – differences in political views, over what policies to advance? Or something else, perhaps deeper and less articulate?

Seeking for answers, a good place to start is the slogan Malm cites: "More ice sheets, fewer bankers". Is this a compelling constellation, and for whom? Are ice sheets commonly associated with bankers, bankers with ice sheets? Of course, in an academic way the constellation has truth to it: the most successful banks/bankers are responsible for driving climate change. Yet this is not what first and foremost shapes most people's opinion about them. For one thing, successful businesses and their CEOs are perceived as admirable in a neoliberal society like ours. Presidents and prime ministers insist they are too big to fail. And if this stance triggers a backlash – as it famously did when Obama assured anxious Wall Street CEOs that "I will help you out" during the finance crisis in 2008 – the anger will be directed at the politician and government in question, not at the banks that expand their power and increase their profits by way of ruining people's lives, forcing them to give up their houses and live in camping wagons or on the street, as happened in the United States. And as for the other half of the constellation, ice sheets are abstract to most people and their everyday life hardships; struggling to pay the bills is eminently concrete. One may respond that in fact the climate emergency is just as "here and now" as is the economic crisis, as brought out in the slogan "No planet, no business", implying that, objectively speaking and according to the science, ecology has primacy over economy, making it absolutely crucial that policy decisions reflect this order of priority.

But they do not, nor are they likely to do so in the foreseeable future. To see why, consider what Malm identifies as the greatest achievement of Extinction Rebellion: the two civil disobedience campaigns of 2019 "massively shifted the point of gravity in domestic politics. They did more to press home the climate emergency than a thousand additional peer-reviewed papers. Towards the end of the year, public concern about the crisis had reached unheard-of levels" (Malm 2021: 129). He adds that "what was perhaps most awesome about XR was the sheer speed of its development" (129).

To be sure, to succeed in "shifting the point of gravity" is a major feat. Five years later, though, we can say that the shift didn't last – and I mean this to apply for the Thunberg-led "Fridays for Future" world-wide demonstrations as well. Both campaigns peaked in 2019. Since then, with few exceptions, things have gone from bad to worse, with Green parties having its share of the vote halved in countries like Germany, France, the Netherlands, and Scandinavia; the youth wings of environmental organizations like Friends of the Earth have seen even bigger losses of membership.

What happened? Perhaps Malm and the scores of others – including myself – who were enthusiastic about the prospects for ever-stronger turnout for protests, with thousands taking to the streets every Friday, and the breakthrough this mobilization would achieve among politicians and governments, should have been more suspicious of "the sheer speed" with which the campaigns in 2019 took off. What goes up, must come down; what goes up very quickly, may go down very quickly. The phenomenon is not new: fighting politically for environmental causes has invariably been at the mercy of other causes not being perceived as *more* important, *more* urgent. Only when other things – it's the economy, stupid – appear to be in order, or fairly under control, has it been possible, in affluent liberal societies like the ones we talk about in Europe, to grant priority to environmental issues. If the relatively rare situations – the year 2019 again – where such mobilization occurs, with widely shared demands for policies to change, *fail* to bring that change, the result is disappointment – and the higher the initial hopes and ambitions, the deeper the disappointment, taking the form of anger in some people, leading to disillusionment, resignation, and apathy in others. We examined this dynamic, overtly psychological yet with important political implications, when discussing the stories of individuals like Sally and Eirik in Chapter 2. We now see that the same dynamic applies to whole groups of people, whereby members' differing ways of coping with disappointment may be a factor in weakening or even dissolving the group as a politically efficient agent: while some resort to therapy, others drop out completely, and others still decide to replace the tried-out methods with more militant ones, stepping up the fight in a way that – we just saw – partly triggers, partly justifies more extreme measures on the part of the authorities: politicians conveniently leave it to the police to take action.

In my view, what is needed for change to happen is that *the burden of argument* shift: it needs to shift not only from one party of the debate or the conflict to the other; it needs to shift *substantially*, with respect to how the practice in question is being perceived – as something good and worthy of being continued, or as something problematic that needs to change. Such a shift would entail a shift in what is regarded, and acted upon, as the default position: the position that is taken for granted as valid, as shared with others, as in no need of justification, of engaging in argument; as to be trusted as allowing for flow, sustaining a sense that what I think and do is both accepted and understood by others: what I think and do equals what *we* think and do. Call it *doxa*, call it "das Man" (Heidegger).

A good example is Greta Thunberg's questioning of the practice of flying. In an affluent country like Sweden, flying – be it for job-related purposes, be it for holidays – is perfectly normal; it is what we do, provided the distance calls for it. In telling my family, my colleagues, or my neighbours that I will be flying when going to a conference in Berlin, no eyebrows are raised; the sheer *commonness* – normality – of this decision guarantees its social acceptance, with no questions, let alone objections, being raised. In pointing out that flying is, by far, the most environmentally harmful thing we do, both in absolute terms and when compared to the alternative ways of travelling,

Thunberg is turning the tables: something that is viewed as perfectly acceptable *in being perfectly normal* is put into question. The innocence is challenged, from now on you cannot simply fly, with no questions asked, no answers needing to be given. You are confronted with your choice, considered as a choice that should and could be a different one, that, as it stands, is problematic, contestable, even wrong. Being forced to justify it, highly unusual as that is, and lacking role models, how do you do it? Yet this *how* question is not the deeper one, only its manifestation and its consequence. The deeper issue is *that* you have to justify it: to enforce such justification, to make it mandatory, is the practical work of the shift.

Return now to the comparison between slavery and climate change that Malm, like so many others before him, invokes in his discussion of historical examples that shed light on how such a shift may come about. In asking what it took for slavery to come to an end, part of the answer is that its critics managed to enforce a shift in the burden of argument: slavery went from being what we do, how things are, perfectly legal at that, to being considered as *something we should not go on doing*, being involved in, profiting from, approving of, and thereby help sustain into the future, having inherited it from the generations preceding us as a crucial element of who we are and what we do. Of course, this "we" is no all-inclusive such, even though both ideology and psychology play a role in persuading the slaves themselves to accept the practice whose victims they are, accept it by way of adapting to it is a normality that is "God-given", inevitable, necessary, allowing of no alternatives; in short, a social order that gives everyone what they deserve, matching their "natural" position and true to the differences in "worth" between whites and blacks according to hegemonic notions of race.

As I mentioned above, Malm highlights "the very notion of slavery as unethical" as decisive to the shift that ended it. To come to perceive something as unethical, as not – or no longer – being on the right side of morality is crucial to accepting that it needs to be questioned, that its persistence cannot be taken for granted, that a change is required. History teaches us that what is nowadays termed "social tipping points" typically have a shift in moral perception at their heart: coming to consider as unethical what used to be taken for granted as morally unproblematic is a prime mover. Be it gradually, be it more or less overnight, after a period of latency, those who defend the existing order need to explain themselves; they need to actively *justify* why that order merits being continued, why it is "better" than alternatives to it now put forward, making for a *contrast* to it that has not previously been on the table, yet from engagement with which there is now no escape. The burden of proof and argument has shifted, and there is no going back, no way of pretending it's not happening, unless you accept being on the wrong side of history.

So much for the ending of slavery. What about ending fossil fuels? Malm cites a climate science report from 2019 concluding that reaching the Paris 1,

5 degrees limit still remains "technically possible" on two conditions: first, to institute "a global prohibition on all new CO2-emitting devices". To ensure that this comes about, says Malm, the climate movement "of millions" should

> announce and enforce the prohibition. Damage and destroy new CO2-emitting devices. Put them out of commission, pick them apart, demolish them, burn them, blow them up. Let the capitalists who keep on investing in the fire know that their properties will be trashed.
>
> (2021: 67)

The second condition for staying below 1, 5 degrees Celsius is

> substantial reductions in the historical lifetimes of fossil fuel infrastructure (68). In short, assets would have to be stranded. Malm comments that no state seems prepared to do this, because "capitalist property has the status of the ultimate sacred realm.
>
> (68)

He goes on to advance the claim I cited above:

> Property does not stand above the earth; there is no technical or natural or divine law that makes it inviolable in this emergency. If states cannot on their own initiative open up the fences, others will have to do it for them. Or property will cost us the earth.
>
> (68f.)

As of this writing, five years later, we can safely say that the conditions are not being met. The momentum that clearly inspired Malm when he wrote his book – the massive turnout for the XR demonstrations in London in April 2019, on the heels of the world-wide school strikes led by Thunberg one month before – is all but gone. The backlash is enormous, fossil-related profits reaching an all-time high (partly thanks to Russia's invasion of Ukraine in February 2022), with fossil-pushing parties winning elections on all continents, "drill, baby, drill" proving a recipe for success not only in countries like the United States and Argentina, but in Norway as well.

What happened? Where did the movement go wrong? Is there something that Malm overlooked, or misjudged, in his analysis?

4.5 Consumers or citizens?

Malm's claim that "capitalist property has the status of the ultimate sacred realm" is still valid, and nothing has happened to prove it wrong. He is right, too, in pointing out that the widespread targeting of the consumer deflects attention away from production (2016: 365), as mentioned above. Granting primacy to property and production is what Marxists do and have always done – primacy in terms of accounting for power relations and responsibility, that is, the class-related asymmetry at work in the economic, political, and moral domains.

Though correct in theory, the primacy of the constellation property/production over consumption is not borne out in practice. Not all of us own private property in the sense important to Marxism; and capitalists would most likely shift investment from fossil to renewables if their profits would benefit from doing so. They are not ideologically or dogmatically committed to the continuation of the fossil option – only economically (commercially) so, as long as destroying the earth brings bigger profits than protecting it.

The component of consumption goes deeper. It does so quantitatively as well as qualitatively. First and foremost: we are all consumers, capitalists, and socialists alike, rich and poor alike, albeit – increasingly – asymmetrically so. Production doesn't *tie* the person to the object (commodity); consumption does, functioning like a glue, an adhesive – if not to *that* particular object (if they weren't exchangeable and replaceable, we wouldn't go on buying new ones, never becoming satisfied), then to the relationship with objects as such. To ensure this is the work of advertising, as Jean Baudrillard showed in *The Consumer Society*, written in the late 1960s, long before advertising was able to expand enormously in time and space by going digital, reaching everyone, everywhere. *Second*, our being consumers in a world where 80 per cent of the energy mix is fossil, just like it was in 1992 when the aim to reduce that amount was first made top of the political agenda, means that most of what we buy whenever we shop is fully integrated into that fossil energy regime, helping deepen and sustain it. Plastic – to pick an obvious example – is virtually everywhere, not only on the supermarket shelves, but in our clothes, our children's toys, our food, in the seas, in the animals, in our bodies, even in our brains. And plastics of all sorts are the oil product *par excellence*, ubiquitous in presence and impact; a long-lasting stuff in a short-sighted society.

At times Malm acknowledges this point – interestingly, more pointedly in *Fossil Capital* than in *How to Blow Up a Pipeline*. He quotes a professor in the Department of Fuel Technology at the Royal College of Science, speaking in 1912: "And to all of us the sitting around the coal fire is one of the most cherished features of our home-life. In abolishing it we might save coal, *but we should lose England*" (2016: 363; Malm's added italics). The implication is clear: "So inextricably bound up with the fire had the English subject become that its extinction would have deprived her of her being" (363). But Malm hastens to add that there can be no one-description-fits-all on this topic – class is the factor that makes the biggest difference, as Marx never tired of insisting. Malm's conclusion thus goes like this:

> The subjects most thoroughly constituted by fossil use-values and therefore resistant to climate change mitigation are the richest consumers. Someone poor, who might pay her utility bill but never flies to the beach, would have far less of her subjectivity invested in the stock and little if anything to lose from a transition.
>
> (365)

Is this convincing? Is it true that the poor person has less to lose than the rich? Or isn't it so that, in case of a transition originally resisted by the rich, being rich would help that person cope better with the change than the poor? Be that as it may, the heart of the matter lies elsewhere. The point to make against Malm's conclusion is that, in today's world, the triumph of capitalism as a way of life – economically and mentally – and as such the only game in town sees to it that, be you rich or be you poor, and regardless of country or region, the life people *aspire to* is that of a successful consumer. Individuals who are looked upon as "losers", writes Zygmunt Bauman, and who look upon themselves that way, are "failed consumers"; failed consumers are failed human beings, and they have only themselves to blame (2007, 2008, 2011).

To be sure, this is a pessimistic point to make against Malm's conclusion, following as it does from his not very optimistic analysis. There is a scene towards the end of the movie inspired by Malm's book on activism, carrying its title, that is telling here. The movie is set in the United States, and one of the protagonists, a young student who takes part in the somewhat amateurish attempt to blow up an oil pipeline encounters one of the men who works on the site. There is a verbal confrontation between the young activist and the middle-aged worker, and it escalates into a fight quickly decided in favour of the muscular worker. He may have lost the argument about the sustainability of oil and the ethics of being employed by a big fossil company intent on "destroying the Earth" (Malm), yet he emerged victorious from the physical fight. Of course, such might doesn't make right; the argument that was interrupted by the worker's anger stands.

What does this confrontation tell us? According to the Marxist formula, left-leaning activists and low-paid workers for big fossil companies objectively speaking are allies: they have the same enemy, the capitalist class who earns hundred times more than the workers and who in doing so destroys the earth. Even so, the activist and the worker perceive each other as enemies, not allies. The worker doesn't feel threatened by the system, but by those who try to bring it down.

Of course, there is a lot to be said about how the encounter between two such individuals is being framed these days: the narrative that determines perceiving the other guy as an enemy, pitting students (city-dwellers, intellectuals) against blue collar workers, the "arrogant" against the "uninformed"

(Hillary Clinton's "deplorables"), and so on. Especially in current circumstances, the task of having ideologically construed misperceptions of who constitutes enemies and who allies replaced by "correct" ones according to Marxist analysis is a an enormously tall order.

What makes the analogy between ending slavery and ending the political-economic regime of fossil fuels so problematic is that whereas ending slavery was about *expanding* freedoms and rights, rendering them truly universal as opposed to selective and particularistic, ending fossil fuels and the kind of lifestyle it facilitates will have to be about *restricting* freedoms and rights, if not per se, then the way in which they are currently inextricably tied to consumption: the sense of *entitlement to consumption*, to increasing it, never reducing it, spans the rich as well as the poor, in the richest as well as the poorest countries on earth. In short, the sense of consumption-intensive entitlement that "objectively" and historically speaking is justified among those on the bottom, those who so far have been excluded from socioeconomic progress, is just as strongly felt, and fought for, by those on the top. Privileges are stubbornly conceived and held on to as well-deserved, not given up lightly, no matter how excessive, nay outrageous they are objectively speaking and when compared to the lifestyle of those worst off.

Let us return to the worker who knocks down the activist, having no patience with whatever argument – political, ecological, moral – he makes for blowing up the pipeline that provides the worker with a living. Is their "conflict of interest" no such thing, objectively understood? Is it a matter of false consciousness – on the part of the worker, at that? Will a more suitable setting for their encounter enable them to overcome their prejudices and stereotypes about each other, coming to see that they are not really on opposite sides of the conflict, but allies against capital?

I doubt it. In Norway, one of the richest countries in the world, where class-based inequalities of income and wealth are modest compared to other nations, workers feel entitled to earn more and spend more than they currently do – having their fair share of economic growth, viewing continued growth as desirable not only for themselves but for society as a whole. Labour unions – very powerful in a country were the vast majority of workers, across industries, are organized – routinely advance demands for higher wages when they negotiate with their employer counterparts. This aspiration also holds for so-called "developing countries" like China and India, which together make up one-third of the current world population. Presidents and prime ministers come and go, but they all – more or less credulously – promise workers that they will get their well-deserved increased part of growing the economy. The worse-off have the right to live like the better-off; comparatively worse-off countries have the right to the same level of prosperity that countries like the United States and in Europe have enjoyed for centuries, often as a result of their colonial (pre)history. This argument has been advanced, successfully, at COP after COP ever since the first summit in Rio de Janeiro in 1992, again and again preventing the deal sought in the name of sustainability: we – China, India, Brazil, Indonesia, and Nigeria – have the moral right to enjoy the same standard of living as the richest countries in the world; and fossil-fuelled growth is the way to attain it – and so far solar and wind supplement and come *in addition* to the fossil energies; renewables do not *replace* them, as they must (see Malm and Carton 2024: 191). Those who deny us that right are hypocrites; they employ double-standards, put "sustainability" and other fashionable buzzwords before them to hide their self-interest and their policies of discrimination. Add to this that in China and India, in particular, the political elite's promise to continue lifting millions out of poverty, enabling an ever greater proportion of the population to benefit from the level of prosperity attained by persistent growth, entering a consumerist lifestyle comparable to the most affluent Western countries, is regarded as a "guarantee" that political stability will prevail and that only fringe groups – intellectuals, students; indigenous groups – will actively seek to topple the regime.

The problem for Marxists like Malm, and for environmentalists who may or may not be thus ideologically committed, is that "progress" is the hegemonic narrative, wed to the regime of fossil capitalism that is the only game in town (with few and marginal exceptions). Whereas young activists are being sent to jail for protesting against a fossil-driven future they see as catastrophic, to be halted by any – or almost any – means necessary, the majority of the population, across ever-increasing class divisions, look upon the fossil-based future and the prosperity it promises as highly desirable, as delivering the goods they feel entitled to. Hell for one party, the young activists, heaven for the other. The contrast could not have been greater. Granted, I exaggerate to make the point – but only slightly.

So what kind of politics is Malm proposing? It is clear from his books – and most explicitly so from his most recent one, *Overshoot* – that he does not think that the change needed will be top-down. In fact the most powerful players in politics and business are more keen on actively preventing the shift away from fossil fuels than in facilitating it. If change isn't going to happen top-down, will it happen bottom-up?

In raising this question, several blind spots in Malm's analysis become visible. As a Marxist for whom the relationship between capital and labour is key to explaining conflict, Malm has surprisingly little to say about the role of labour in the current crisis. Whose side is labour – ordinary workers and their unions – on? Can the antagonism and the mutual suspicion that played itself out in the fight between the worker and the activist above be generalized so as to illustrate how these parties tend to view each other? If they do tend to see each other as enemies rather than allies, what follows for the fight against fossil capitalism? To the extent that labour is critical of the power exerted by capital, is it because "capital destroys the Earth", that is to say, for ecological reasons? Or is it because the goods produced are distributed in a grotesquely unjust way, making the rich richer and the poor poorer when there is "objectively" enough for everybody to enjoy prosperous lives based on equality and fairness? Of course, it will be pointed out that this juxtaposition between the possible critiques is based on a misunderstanding: a just distribution presupposes a sustainable relationship between society and nature, economy and ecology; you cannot have one without the other. This may well be true. But it is the answer given by academics, not the majority of workers, and especially not those employed in the fossil industry itself.

Malm's silence on the role of the working class is conspicuous. Has he chosen not to address it because he realizes how complex it is, given that Marxist assumptions about labour as the revolutionary subject have been widely regarded either as simply mistaken or, at least, as questionable, ever since the rise of fascism and Nazism? The first generation of the Frankfurt School – Horkheimer, Adorno, Benjamin, Marcuse, and Fromm – spent decades trying to explain why large portions of the working classes were attracted by the political forces that sided with capital, helping capitalism to maintain its hegemony, culturally and economically. Today, of course, the roots and manifestations of the crisis are multiple in a way not seen before, nature having become the locus of disruptions and degradations previous generations of Marxists could not anticipate, being trained to conceive and criticize the costs of capital's rule in terms of *human* (social, class-based) suffering and injustice, not in terms of devastating consequences for the natural world.

Given this perspective, a remark made by Malm at the end of *Overshoot* is striking. Commenting on the election of Gustavo Petro as Colombia's first-ever left-wing President and his promise to phase-out fossil fuels, Malm writes that "if the ban on [oil] exploration withstood the immense pressures from bourgeoisies local and global, Colombia would be on track to have zero output by 2034 – the task *rich* countries should have shouldered" (2024: 257). He assumes, or even takes it for granted, that if there is opposition to the phasing-out of fossil fuels, it will come from "bourgeoisies" local and global, not (also) from the working class. Whether or not this holds for the situation in Colombia, I cannot say. But the general picture is that political parties promising to continue fossil-based economic growth attract more votes among workers, and win more elections, than parties promising to take steps to end the era of fossil fuels – for the most part not by radical steps at that, but modest one, so as not risk losing support among those committed to the fossil regime and the growth it facilitates; mentally and culturally committed in the form of the sense of entitlement explored in earlier chapters, and not necessarily by way of earning one's living in the industry. Kamala Harris's election campaign turnaround in 2024 on the issue of fracking is a case in point, more representative, I think, than Malm's comment about Colombia. Afraid to alienate working class voters in Pennsylvania, Harris declared that she had changed her mind about fracking and would see to it that it continue were she elected President. Thus the competition between her and Trump with regard to securing the support of workers was a competition about who of the candidates promised most steadfastly to *continue* – not end – fossil industries in the years to come.

For lack of the working class playing an important part in the fight against fossil capital, who does, according to Malm? This may appear an unnecessary question to ask, given his account of political movements using civil disobedience, and his highlighting of the so-called "radical flank". Important though the victories brought by this kind of activism may be, their achievements are rarely permanent and there have been a number of setbacks the last couple of years. What is needed to bring about change of the magnitude required?

Absent a collective subject – be it the working class (as if it were a homogeneous entity – it isn't) – attention has to shift from extra-parliamentary opposition such as Extinction Rebellion and Fridays for Future to the role of the state and of institutions. But again Malm has little to offer. Even in his most recent book *Overshoot*, coauthored with Wim Carton, there is no systematic account of either. And to the extent that they are part of the analysis, their role is that of bodies intent on preventing, or postponing, or watering down, the change needed, just like the Harris example above goes to show. The state, and the governments that come and go in so-called liberal democracies, are part of the problem, not the solution. What they guarantee – to the voters, and to the fossil fuel companies – is that nothing of any consequence will change, especially not something that threatens economic growth. Business as usual is and remains the default position, making the imperative to overcome it both more urgent than ever and more difficult.

This being the situation, Malm's skepticism about the potential for mobilizing the state and its institutions to shift from the wrong to the right side of history, given the climate crisis, seems well-founded. By implication, his assessment concurs with that offered by Judith Butler and other leftist critics of neoliberalism, discussed above: the state is to be mistrusted, its institutions are so many vehicles of oppression and exploitation, not only of manual labour as in Marx' time, but as human capital trained to exploit itself ever more ruthlessly on the level of the individual, turning them into competitors where group-based solidarity is more needed than ever.

I accept that there are good reasons to view the state in this critical light. Yet giving up on it as force necessary to bring about and help sustain necessary change is not a good idea – beware of throwing the baby out with the bath water. As I told Malm when we discussed this at a conference in Oslo in May 2022, the only agents his analysis seriously reckons with are fossil companies and activists trying to bring them down. It is as if all the other agents that political science pays attention to in describing conflict – who defends the old order and who seeks to topple it, namely political parties, the civil society, the institutions of the three branches of power, are bracketed, leaving the confrontation and the fight itself to activists trying to blow up an oil pipeline. Of course, to put it like this is a simplification. But the point stands: the state is pictured as a passive onlooker or bystander, stepping in to protect the fossil companies if need be, increasingly nowadays (as we saw) by criminalizing activism. Malm's response was that what really ought to be criminalized is the production of fossil fuels; that it should be seen as a form of violence; that this form of money-making kills; that efforts of resistance, meant to prevent death in so many forms, must be regarded as legitimate self-defence; and that, as the crisis gets worse (and it will), the destruction of property and fossil infrastructure is "an idea whose time has come", such that "people will learn to take things into their own hands".

The upshot is that the struggle against fossil fuels is one of David – or so many distinct Davids – against Goliath. The sort of popular support for taking on Goliath and declaring the persistence of fossil fuels "criminal", a violence that needs to stop, has not materialized in recent years; instead the opposite trend has been very much on the offensive, as Malm himself fully acknowledges in *Overshoot*.

The reasons why the kind of mobilization Malm envisioned a couple of years ago has failed to happen are too comprehensive to address here. What we can say, again on the negative side, if you like, is that *without* being anchored in major institutions, and for lack of enduring efforts made to bring about such anchoring, the sort of movements Malm pins (pinned?) his political hope on – besides XR and Just Stop Oil, Black Lives Matter may be mentioned – risk being short-lived; after a period of gaining momentum, of taking "everyone" by surprise by how rapidly and massively the turnout explodes, the timing seems right, and the *Zeitgeist* is on their side – they either slowly, or just as quickly, fade, with support dwindling and media-coverage going from front-page to the margins, because other things happen – a pandemic, a different sort of crisis – and people get either exhausted or depressed or simply otherwise preoccupied.

This is what I had in mind when I said, above, that what goes up quickly, tends to go down quickly, as applied to groups engaged in resistance. Changes of the magnitude needed in the climate emergency need to be *institutionalized* in order to bring about and translate into practice the change needed: they cannot, if they are to mark a real change, remain *outside* the institutional bodies and structures. So, indeed, it does seem a good idea (as Malm suggests) to *prohibit* the production of "Earth-killing" fossil fuels – perhaps even the only way to make sure that it happens. But deciding to do so, and seeing to it that it is enshrined into law, hence sanctioned if the relevant agents (companies) fail to abide, is the work of institutions. To be sure, they need to be pressed to get into action, on this side of the conflict, and such pressure – history teaches – will often be the part played by groups of resistance outside the established bodies of power. Yet, to be successful, at some point the struggle for change will need to see institutions as allies, as partners in change, not as the enemy – and vice versa. Having helped reach a social tipping point in the mid-1960s, the US civil rights movement is a case in point: it achieved being heard by the President, who pushed new legislation, and the rest is history. Will there be something similar in the case of the climate emergency? Say, a "George Floyd moment", to allude to the momentum gained by Black Lives Matter? Can policeman Derek Chauvin's killing of a black man unable to fight back be compared to how profit made from fossil fuels kill?

I find it difficult, if not impossible, to answer questions involving the latter sort of comparison, suggested by Malm in our discussion. I find the comparison with the ending of slavery, explored above, more instructive, allowing for a clear answer: namely that ending fossil fuels will prove much harder, not only, or even primarily, because the adversary is so mighty, but because fossil-based capitalism is a cultural as well as political order that the majority of people – rich or poor – look upon favourably, as bringing more prosperity than any rival regime on the horizon. Arlie Hochschild's (2016) finding in Louisiana, discussed at the end of Chapter 3, that people suffering the negative, even lethal consequences of the fossil industry come forward to declare that, on balance, "it is worth it", speaks to this phenomenon in a way replicated, in similar circumstances, around the world. As we have seen, the investment in the dominant lifestyle, whatever its adverse "side effects", is not a matter of rationality or logic or emotionally disinterested and detached cost–benefit analysis on the part of those affected; it is a stubborn, recalcitrant investment, resilient to change in its own way, anchored in a deeper level, psychically and culturally, than that of the aforementioned procedures and criteria, to put in a Freudian manner. If there is at all an element of calculating and summing up the pluses and the minuses, it is of the form: having invested so much in helping maintain and reproduce the current order, I cannot suddenly change my mind – I cannot turn against it, for in doing so I would turn against myself, and against us. "Pollution is the sacrifice we make for capitalism", says Jackie to Hochschild (2016: 179), the message being: "and it is worth it". Indeed, as we saw in Hochschild's interviews, the longer and more stubbornly, even bravely, people have clung onto their investment – psychologically and emotionally, ideologically and economically – as a part of a fossil-based society wedded to growth, to its promise of prosperity and progress, the less likely that they will change their mind, shifting from focusing on its hoped-for goodness to recognizing its increasingly abundant and visible badness. For that change of heart and mind, *too much* is at stake. This is the insight Malm alludes to in his quote, brought above, that to give up coal would be to "lose England", that is, the essence of who and what we are.

Notwithstanding the validity of this insight, there is perhaps a more straightforward, less subtle and psychically anchored aspect to the potential for resistance and mobilization. In *Overshoot*, Malm offers an argument in favour of the primacy of production over consumption that I problematized above, an argument that improves on the one he made in *How to Blow Up a Pipeline*:

> From Kyoto to Paris and beyond, climate politics has been woefully fixated on the demand side, when focusing on supply makes so much more sense: the points of production are fewer and more easily shut down than those of consumption; the problem would be addressed at the source. But perhaps most importantly, supply side politics *emanates from local resistance*, from which it works its way up to the higher scales. Moreover, insofar as they succeed, blockades and other types of place-based resistance *count as politics in and of themselves*, on a par with anything a government might legislate.
>
> (Malm and Carton 2024: 243)

Having discussed various historical experiences with organized forms of activism and how they have been met by the authorities, I now turn to the role played by individual agents who have sought to mobilize wider support for the structural changes needed in the face of environmental crisis. In doing so, I shall explore a topic that Malm, like so many others engaging in "the politics of climate change" (to cite sociologist Anthony Giddens' book title) fails to grant proper attention, namely the part played by whistleblowing and leadership. We shall see that the responses to acts of whistleblowing, and the obstacles to leadership of the sort deserving the name in times of crisis, shed important light on the conditions under which resistance currently struggles.

4.6 Whistleblowers: Carson and Thunberg

Silent Spring by Rachel Carson, a relatively unknown marine biologist without a university position when the book was published in 1962, is widely credited with alerting the world to the hazards of environmental poisoning (in particular, pesticides like DDT) and remains an inspiring classic to "green" movements around the world to this day.

I shall not present the findings in Carson's book here nor the specifics of the critique she levelled at the powerful pesticide industry and its many allies in the United States. What makes Carson relevant to my discussion is the way in which her intervention, and the responses it triggered, help us understand the phenomenon of *whistleblowing*, crucial – I shall argue – to resistance.

In her monumental biography of Carson, *Witness for Nature*, Linda Lear succinctly sums up the reaction to Caron's critical analysis of the environmental impact of pesticides in the following two paragraphs:

> The pesticide industry trade group National Agricultural Chemicals Association (NACA) spent well over $ 250,000 in its efforts to persuade the public of Carson's errors and to protect their threatened interests. They were worried about the possibility of new government regulations, expensive testing procedures, and stricter federal state oversight. As a result, they sought to discredit Carson as a scientist by ridiculing her evidence as well as her conclusions. By denigrating Rachel as nothing more than an emotional female alarmist, they hoped to win the public relations battle in the marketplace and avoid a legislative battle in Congress.

But many in science, government and industry understood that the essential issue that Carson had raised was not one that could be won or lost by appeals to the consumer. They recognized *Silent Spring* for what it was: a fundamental social critique of a gospel of technological progress. Carson had attacked the integrity of the scientific establishment, its moral leadership, and its direction of society. Holding up before them their resolute carelessness of the natural world, she dared to make their sins public. The fury with which they attacked her reflected the accuracy of her moral charges. Since her facts were essentially irrefutable, they disparaged Carson as the agent of a message that had to be suppressed" (Lear 2009: 428f.).

This being the mood, it was inevitable that the repertoire of accusations against Carson would include the proverbial "she is probably a communist", as well as unstoppable attention on Carson's gender and her being "a spinster with no children", as were it suspicious by itself. Her arguments, critics would reiterate, were "exaggerations born of hysteria at worst and an overly sensitive nature at best"; arguments "more emotional than accurate" (2009: 430, 461).

So much for Rachel Carson, whose legacy has proved more enduring than that of her critics and detractors taken together. Carson was very much alone when she waged war on an immensely powerful industry and its allies in business and politics, as well as in academia, it would turn out. For lack of the backing of an institution and the kind of support and protection it may bring when somebody chooses to speak truth to power, Carson had to rest her case on her credentials as a scientist without tenure; on the merits of her arguments, and last but not least, on the integrity that her independence granted her as its one decisive asset over against her well-connected critics.

When Greta Thunberg at the age of 15 decided to sit down with her hand-written poster "Skolstrejk för klimatet" ("school strike for the climate") on a warm August day in 2018, in front of the parliament in Stockholm, she did so as a completely unknown child with no connections whatsoever, indeed, without even the backing of her own parents. Almost to the day, thirty years had passed since that also very warm day in 1988 in Washington, DC, when the respected and NASA-affiliated physicist James Hansen gave his testimony in Congress, telling the lawmakers that global warming was happening, that it was caused by anthropogenic emissions of carbon dioxide, and that unless halted, the consequences for human societies would be severe. Thunberg may or may not have known about Carson and Hansen when she decided to embark on her "school strike" as its sole participant.

Thunberg's personal story, with her autism diagnosis and eating disorders, is well known and needs no repeating here. At the age of 11, she became aware of the impact of climate change on wildlife in the Arctic, and immersed herself in whatever books and films she could come across that would deepen her understanding of what was going on. From early on, she was able to grasp the essentials of climate science and the warnings issued by the IPCC, growing more dire over the years, and largely falling on deaf ears as far as the required changes of policies are concerned. Thunberg did what others – be they grown-ups or children – did not, and told each other not to do: she took the climate emergency personally, suffering it as an existential not technical or distant issue, yet understood that the leverage of power needed to "solve" it is genuinely political; those with most power carry the responsibility for averting disaster and for securing the safe future children everywhere are entitled to.

The combination of knowledge-based hard-nosed realism and uncompromising resolve that soon came to define her public persona was in display already on the eve of her debut as a lone protester. A conversation with her father Svante, who in his soft-spoken manner was trying to talk Greta out of her plans for a strike, goes like this:

> "The journalists will ask you all kinds of questions", Svante said to Greta. For example: "Is it your parents who have persuaded you to do this?"
>
> Greta answers: "In that case I will tell it like it is. That in fact I have been putting pressure on you, not the other way around".
>
> "Exactly", responds Svante.
>
> "All right, then", says Greta in a slow thoughtful way.
>
> But there are no solutions within the current system. The only thing we can do, is to begin to treat the crisis as a crisis. [...] To all those who do not understand how serious the situation is, going on school strike for the climate will appear incomprehensible

> says Greta, in an upbeat, almost euphoric way. "And since nobody knows about it, hardly anyone will understand. I am going to be incredibly hated for it", she laughs.
>
> "But maybe the children will understand it", suggests Svante.
>
> "No. Children act like their parents do", she replies. "I haven't met a single child who cares about the climate. Everybody says that the children are going to save us, but I don't think so".
>
> (Ernman 2019: 233ff.; my translation)

The reluctance of Greta's father Svante to her "fix idea" of starting a school strike all on her own is easy to understand for parents everywhere. They want to protect their child from being ridiculed, made into a laughing-stock and perhaps downright hated. Svante reckoned that in going public with her anxiety and desperation over the climate emergency, Greta would lose out, be disappointed and come to regret her action. He warned her to protect her.

Yet, and in keeping with what we have seen in Norgaard's and Lertzman's work in Chapter 2, parents' endeavour to protect their children is often a concealed attempt on the part of grown-ups to protect themselves (see Vetlesen and Willig 2018). The brutal truth from which they seek to shield their children, is a reality "out there" they themselves do not wish to confront. Greta's terse remark that "children act like their parents do" is a reminder of the obvious fact that grown-ups are the ones who embody authority and serve as role models to their children, not the other way around. This makes for the basic asymmetry that obtains between the generations and that cannot be done away with by pretending either that adults and children enjoy a symmetric relationship with equal power and responsibility, or – worse still – that children should be role models for their parents. Says Thunberg, having just turned sixteen in February 2019, and having in the course of six months expanded her audience from her family members to the entire world: "I agree. I am too young to do this. We children shouldn't have to do this. But since almost no one is doing anything, and our very future is at risk, we feel like we have to continue". Thunberg minces no words: "Your generation is failing us. But the young people are starting to understand your betrayal. The eyes of all future generations are upon you. And if you choose to fail us I say we will never forgive you" (2019: 31, 99).

Thunberg's most iconic speech is probably her address to the UN General Assembly in September 2019, "How dare you?". She makes it clear that the fact she has been invited to speak in such a prominent forum is no proof that "progress" is being made; she is not flattered, let alone reassured by having been given the opportunity to speak her mind in front of prime ministers and top diplomates from all over the world. To the contrary: "This is all wrong!" is her message. In her characteristic shrewd and no-nonsense manner, she refuses to accept the role that so many powerful adults are attributing to her, namely to enact "reverse parenting", whereby the child steps in and does what the grown-ups should have done long ago. Time and time again she has vigorously opposed such role-reversal, complaining about how slow and reluctant the supposedly "adults in the room" are to get it. She is particularly scornful, almost sarcastic, about the hypocrisy at display when the "most powerful leaders" in the world – be it Obama, or Merkel, or Macron – express "hope" when witnessing how inspiring Greta evidently is to millions of young people. "When your house is on fire", she says,

> you don't sit down at the kitchen table, full of hope, and talk about how fine the house will be when you rebuild it after the fire has destroyed it; to the contrary, you call for emergency and exit the house as soon as possible.
>
> (Ernman 2019: 163)

Yet, for all its rhetorical power, the analogy is misleading, as she knows only too well: the planet is not a house we can leave, trusting others to repair it and then come back and begin afresh in a new one.

Looking back on Thunberg's kind of activism, three factors stand out. First, the fact that she commented on, saw through and resisted the countless attempts to put her into the role of the "responsible" adult in the form of reverse parenting, leading the way as it were not only by way of example to other children, but to adults as well, has proved a crucial asset for her. She has not sold out; she has remained uncompromising, never losing sight of what is important – not words and pledges, but action aiming for radical change: "All that will remain of our political leaders' legacy is that they will be remembered as the greatest villains of all time, because they have chosen not to listen [to the science] and not to act" (2019: 35). In this way, Thunberg has kept on doing what propelled her to her initial breakthrough: insisting that the emperor is naked, and that pretending otherwise is a non-starter.

Second, by articulating the unease and anxiety felt by many young people toward the future, Thunberg has disproved the common assumption in climate psychology à la Stoknes that I criticized in Chapter 2, that expressing fear and other "negative" emotions will most likely lead to withdrawal and apathy, and therefore prevent political mobilization from happening – the assumption, that is, that mobilization for change is impossible without a message of hope, based on a shared optimism about the future. Thunberg's way of speaking about the crisis "as the crisis that it is", without excusing herself for not subscribing to the mandatory imperative of optimism, proved to inspire more people to taking to the streets than all the speeches ending on a note of hope taken together. Contra Stoknes, then, a message about having a nightmare may prove no less politically potent and inspirational than Martin Luther King's classic "I have a dream". In her most emotional speech – "How dare you?" in the UN General Assembly in September 2019 – Thunberg combined expressing her anxiety with giving vent to her anger – she was furious, frowning and crying as she finished. Yet, as with eco-grief, the anxiety and the anger presuppose, as well as resonate with, something else, and hopefully more enduring: love. What she fears for, what she is angry with, is the ongoing relentless destruction of planet Earth that is at the heart of business and politics as usual. Her target is the persons and the forces responsible for this happening – for harming the life that we love, in all its manifestations, encompassing so many other-than-human lives.

Third, Thunberg has been instrumental in seeing to it that the burden of evidence and the burden of argument have shifted. The taken-for-grantedness and presumed innocence with regard to expanding the production of fossil fuels, of simply continuing with the very practices that more and more visibly alter the climate and disrupt the natural world we are part of and dependent upon, is gone. Not Thunberg's pointing out that the future is being stolen from the young is the stance in need of justification and more evidence, but the sleepwalker-like assurance on the part of politicians and CEOs that continuing to grow the economy will help save the planet. Precisely this kind of proclaimed "realism" and "rationality" is the problem, not part of the solution; a reality-denying madness that is no less dangerous for being so obsessively held on to by the powers that be.

Inevitably, Thunberg has had her share of critics, some of whom are worth mentioning here because of what they tell us about current prospects of activism.

For a start, there is the outpouring of hatred that Thunberg has been met with from her early rise to fame. As I mentioned above, to a remarkable extent the hatred mirrors divisions salient and much talked about in our era of "polarization": divisions of gender, class, generation, and geography. This being so, the most vocal and aggressive attacks on Thunberg, mostly but in no way exclusively disseminated on social media, come from men rather than women, individuals fifty-plus rather than young, without higher education, and living in rural areas rather than big cities. The attacks vary from the personal, typically zooming in on Thunberg's well-known autism diagnosis and history with eating disorder, to the ideological, alleging that she is a communist, or brain-washed, or completely ignorant about the "facts"; in short, a confused and troubled child feeling things, believing things, as opposed to knowing things and understanding how the world works. For lack of academic studies of this material (certainly in the pipeline), I must be careful about speculating about the more precise profile of these critics. That said, I assume that a significant portion either works or has worked in the fossil industry. For what it's worth, to expect as much is corroborated by my own thirty years-plus experience as a public intellectual debating the climate issue in Norway.

Though illuminating, a different line of criticism from the person-fixated one is more relevant to my discussion. This brand of criticism is well-informed, published in serious newspapers, and concerned with the politics of Thunberg's activism – what is it aiming for and what it has accomplished. I shall restrict myself to one case in point.

On October 24, 2024, the German left-leaning newspaper *Der Freitag* published an article by Georg Seesslen, a respected author of many books on politics, especially neoliberalism and its dismantling of the welfare state, inviting comparison with scholars like David Harvey and Zygmunt Bauman. Seesslen's article is titled "Das Märchen ist zu Ende", that is, "the fairytale is over". Note that in German, "girl" is "Mädchen". Seesslen is playing with the association between a fairytale being over and a girl's dream of change coming to an end.

What, then, has brought Thunberg's activist struggle for change to an end? Seesslen has a particular decision of hers in mind: her signing a public plea initiated by Naomi Klein for boycotting Israel because of what's happening in Gaza. In Seesslen's view, Thunberg's repeated declarations of "solidarity with Palestine and Gaza" demonstrate how fatal it is to "divide the world into black and white, good and evil, our truth and the lies of the others". He invokes Thunberg's "autism" and her statement that for her, "political engagement has a therapeutic effect", to argue that, far from being a model, Thunberg reveals how "monomania" – staying focused on one big issue – is more likely to lead to blindness than to politically useful insight. Hence the once-cherished "innocence" of the child Greta is over; her moral authority is in shatters, her integrity irredeemably compromised. The dire truth that so many people, adults and children alike, for so long have denied, namely that Greta has served as a "projection", as a "substitute object" for various people's "inner conflicts", is finally out in the open (2024: 21). We now realize that her "childish fits of anger" have been of no help whatsoever in tackling the problem of climate change.

In some respects, Seesslen's criticisms are unmistakably German: politically, the tolerance for condemning Israel's attacks on the population of Gaza, calling them "genocide", is lower in Germany than in most other Western countries, especially on the political left, and for historical reasons having to do with the Holocaust and so with post-World War II Germany committing itself, as a nation and per identity, to the unconditional protection of Israel and its right to self-defence. Condemning Israel's military operation on Gaza under the Netanyahu government is often regarded as "anti-Semitism". Seesslen is no exception. Second, in no other country has the Thunberg-initiated "Fridays for Future" movement been as big and influential as in Germany, led by Luisa Neubauer. It is well known that, following Thunberg's participation in pro-Palestinian protests against the

Israeli killings of more than 50 000 persons, most of them civilians, in Gaza, the long-time close relationship between Thunberg and Neubauer is strained. Demonstrating just how representative is Seesslen's attack on Thunberg, Germany's most important periodical *Der Spiegel* ran a title story entitled "Irrweg eines Idols", about how an idol has gone astray. Note that the word "Irrweg" carries a double meaning, going wrong and being mad. The message is clear: as far as Germans are concerned, the Thunberg era is over.

I cannot pursue the German debate here. But Seesslen's article is illuminating beyond that particular context.

To begin with, a striking feature of this line of attack is the way it mixes the personal and the political, in effect making points of political disagreement by way of *ad hominem* characterizations. Presumably, Seesslen would not have felt so free to rest his case on this mix were he discussing a respected politician. And if he thinks he can fault "the child" Greta for mixing the personal-emotional with the political-rational, he should abstain from doing just that when critiquing her. Indeed, the vehemence with which he attacks her self-understanding and her activism confirms his point about the dangers of "projection" in a way he does not intend. Similarly, if anger is a non-starter in political debates and in attempts to tackle a crisis like climate change, why is it that Seesslen cannot hold back his anger and stick to his own "adult" advice about sticking to the facts in a nuanced manner? "Reality", he ends his article pointing out, "is dirty, bloody and very complicated", evidently targeting Thunberg's talk about the causes as well as the solution to the climate crisis being "black and white": "we simply need to stop the emissions of greenhouse gases" (2019: 29). To be sure, one can always say that things are complicated. But what is more complicated than Thunberg realizes here: *that* emissions have to be cut, the sooner, the better; or *how* to do it?

It's a pretty safe bet that, from Thunberg's point of view, Seesslen's article is an exemplary case of the genre she was fed up with even before turning an activist at the age of fifteen: an adult engaging in patronizing and bullying a child, calling out the bluff of its (her) innocence and the way anything she can say or do will fail the "reality test" so effectively monopolized and policed by adults-in-the-know; adults who repeat *ad nauseam* how hard it is to change the world, no matter how disastrous the trajectory to which it is locked.

4.7 Autonomy and sacrifice

In my view, Carson and Thunberg are whistleblowers. I now turn to the work of American philosopher C. Fred Alford to provide a theoretical framework enabling us to understand what this means, and what it tells us about the relationship between individual and organization in contemporary society.

Alford's book *Whistleblowers: Broken Lives and Organizational Power* is based on interviews with persons who have chosen to go public with criticism of the institution they are part of.

In the spirit of Freud, and countering popular views about how our liberal society hails the truly "autonomous" individual, Alford argues that the experiences made by people who at some point have chosen to blow the whistle demonstrate "the inability of the organization to tolerate autonomous individuality in the first place; everyone's individuality is a resource to be sacrificed in the name of the group" (2001: 11). He continues:

> To be an individual we must divide ourselves from who we are and what we are doing without falling into ironic detachment. We have to be separate enough from ourselves to have someone to talk to but not so separate that the different internal voices can't hear each other. [...] Individuality means talking with oneself about what one is doing in a spirit serious enough that it may make a difference in what one does. It is the whistleblower's ability to talk to himself about what he is doing that explains, if anything can, why he blows the whistle. [...] Individuality is not a solitary project but a group one, whose medium is culture.
>
> (2001: 12, 13)

We are reminded of the Socratic proposition I cited when discussing Gardiner in Chapter 1: "It would be better for me that...multitudes of men should disagree with me than that I, being one, should be out of harmony with myself and contradict me" (quoted in Arendt 1978: 181). If there ever was a "positive teaching" suggested by Socrates, it is this: better to be true to oneself than to give pride of place to loyalty to others, to the group one belongs to.

Why is this Socratic teaching relevant to whistleblowing? Alford says: "In theory, anyone who speaks out in the name of the public good within the organization is a whistleblower. In practice, the whistleblower is defined by the retaliation he or she receives". He adds: "Somewhere between half and two-thirds of the whistleblowers lose their jobs" (2001: 18); that's how bosses typically respond when an employee goes public – or threatens to go public – with his or her observations of unethical or illegal practices within the organization. In retaliating against the whistleblower – sacking her if she doesn't comply, telling her that nobody wants to hire a whistleblower (which is true, with very few exceptions) – the organization is stating that "there is a certain kind of person it cannot stand in its midst" (20). Initially, the person considering blowing the whistle asks to see the boss with one simple aim: that he be willing to discuss the issue with her, to take her concern seriously, and respond accordingly to deal with the problem so that malpractice will end. Typically, this expectation is not met: the boss is unwilling to discuss the issue that is the heart of the matter; unwilling, also, to talk about not wanting to talk about it. "My subordination was the only issue". Alford comments: "We don't know how important being subordinate is, we don't properly estimate the borders of subservience, until we cross them". For many of those he interviewed, there is something "crazy-making" about the experience, akin to the double bind that was once said to cause schizophrenia: "Don't talk about ethical issues, and don't talk about our not talking about ethical issues" (21). Alford quotes Zygmunt Bauman's (1989: 215) contention in *Modernity and the Holocaust* that "all social organization consists in neutralizing the disruptive and deregulating impact of moral behaviour", adding that this is what he means by saying that "organization quells individuality" (21).

Alford lists five principles that his empirical material suggests are key in whistleblowing. Persons who decide to blow the whistle are not loyal to the principles as such "but to ideal selves who embody these principles"; the principles need to be embodied in oneself, in the sort of self one aspires to be, regardless of whether "multitudes of men should disagree" (Socrates). These are the five principles: an imagination for consequences; a sense of the historical moment, involving a sense of one's responsibility to others; identification with the victim, involving a concern for others, but also, and importantly, a refusal to identify with the aggressor, who is usually in a position of authority; not very good at doubling, which means not being very good at walling off one's moral principles from the way one lives one's life; and finally, shame, in the form of wounded narcissism: the shame many whistleblowers feel is connected to the failure of others to be moral, feeling ashamed for the human race (85).

True to his profile as a psychoanalytically informed philosopher from his first books, Alford links whistleblowing to the phenomenon of narcissism, looking upon it as "narcissism moralized". I shall not dwell on this part of his work here. But Alford's debt to Plato is worth mentioning.

Narcissism, suggests Alford, is a special type of anger, anger understood as akin to what Plato (*Republic* 439e-441d) calls *thumos*: the human spirit that feels indignation at baseness and corruption. "Thumos", explains Alford, rebels at anything that would bring shame upon the self. The trick is to educate thumos so that it gets shamed by the right things, or should I say the wrong ones? It is not an easy task, says Plato, but it has the advantage of working with nature, not against it (78).

Yet "working with nature" in the positive sense intended is not easy. It is a tall order more often sabotaged than facilitated by the organizations persons who consider blowing the whistle belong to. In the spirit of Bauman's *Modernity and the Holocaust*, Alford contends that

> it is the autonomous ethical individual who is the real threat to the organization, and the law finds little room for such people. Or rather, it finds room for them as an idea in such a way as to render the individual behind the idea expendable.
>
> (113)

He faults Habermas for his declared faith in the procedures of Western law, considering them a mirror of what he famously calls "the ideal speech situation" in which only the force of the better argument may prevail. In reality, however, "the rules of civil procedure reflect a system imperative more concerned with preserving the predictability of outcomes than understanding". Indeed, Alford goes on, "the law has hardly a clue about how to protect the individual who resists the organization in the name of the public because our society has hardly a clue about what is public" (113). True enough. Even though Habermas, like Rawls and discussed above, offers a fine theory about civil disobedience and its importance as an "emergency brake" in liberal democracies, real-life whistleblowers, and the retaliation they typically experience, are conspicuously absent in his overly formal and procedure-oriented approach.

"The whistleblower is a political actor in a depoliticized world", Alford finds (113). He concludes his study like this: "To act politically in this depoliticized public space is to be a scapegoat. In this regard, at least, the law is true to its original purpose according to Rene Girard, sacrificing the individual so the group can be safe" (130). (Note that Girard's famous theory doesn't really apply to the cases considered here, because the scapegoat he concentrates on is arbitrary, chosen by lot.) Alford concludes: "The real threat posed by the whistleblower is to remind the organization that it belongs to the larger world. This is what thought represents – a way of thinking whose boundaries are not those of the organization" (130).

I think that Alford's analysis of whistleblowing and the responses it meets help us better understand the experiences made by Carson and Thunberg. Obviously, there is one big difference between their cases and the ones providing Alford with his material: Carson and Thunberg are not members of an organization, and their decision to blow the whistle and sound the alarm to the wider public is not an action taken by an institutional "insider". Indeed, in Carson's case, her lack of a university position was weaponized against her from the start: being an "outsider" to the authorities, corporations, and industry she was criticizing, her findings and assessments were allegedly without scientific merit and credibility, thus not to be taken seriously. In Thunberg's case, it wasn't the school whose pupil she was she was targeting when sounding the alarm – it was "the wider world", to recall Alford's concluding statement: "The real threat posed by the whistleblower is to remind the organization that it belongs to the larger world". This means that neither Carson nor Thunberg could be fired: they could not be silenced by threatening them with losing their jobs and their prospects to get a new one.

Despite this contrast with Alford's sample of whistleblowers, Carson and Thunberg both experienced being individuals punished for their autonomy by being sacrificed "so the group can be safe" – the group being here society at large. This is the standard frontline self-defence manoeuvre to which the groups feeling attacked deploy to neutralize the threat as soon as possible, hoping that for lack of belonging to a group providing protection (though as we have seen, that's not necessarily what groups do), the one defiant alarmist will sooner or later succumb to the loneliness suffered.

And yet, for all the suggestiveness of this line of thinking, history shows again and again that the risks whistleblowing entails for the organization may prove even more consequential for it than for the thrown-out individual. Precisely because of the David against Goliath constellation, the larger world – larger than the organization in question – may rally to support the lone protester, not the mighty powers hell-bent on crushing her, sometimes in broad daylight and with methods that, it turns out, do not tolerate that spotlight. To be sure, to act in a "heroic" manner, to take on forces so much bigger and more resourceful than oneself, will often bring out the worst in people watching it happen, be it their envy, their aggression, their pent-up grievances and frustrations. The retaliations are carried out on behalf of the group whose sense of entitlement has been exposed and challenged by the defiant individual, as in the charge: "Who does she think she is, calling into question our way of life, our values, our identity, our habits? What gives her the right to do this?"

The answer, to the extent that there is one, valid for various cases, is to do with moral authority and moral integrity – virtues we tend to think of as necessarily belonging to the distinct individual; virtues not based on power in a material sense, but on demonstrated willingness to stick to what one believes is right, just and fair, no matter the consequences. To put it like this is, of course, to allude to the role of narcissism, of holding on to an ideal self, that Alford highlights.

It is difficult to admire in another person what I sense to be lacking in myself – this contrast being the stuff out of which envy and vindictiveness are made, from Shakespeare to Freud and Melanie Klein to Trump. It takes maturity to admit that I can, perhaps even should, admire in that person what I do not have, cannot do, cannot be. Yet, by way of example, such a person undeniably shows that it *is* possible to mature, to change, to contribute to what is good – for us as a society – instead of *seeking to destroy it because it is good* and I myself, or the group I belong to, do not possess it, this being the essence of envy according to Klein (1989).

Whistleblowers who make the experience that "the only thing my boss wanted to talk about was my insubordination", as we saw Alford point out, unwittingly challenge a core feature of neoliberal culture and mentality writ large, not confined to the institutional setting, namely the way *loyalty* has come to be understood. As employees in the public sector lucky enough to have been subject to New Public Management reforms during the last three decades or so know very well, the mandatory first question during an interview with the boss is: "When you hear the word 'loyalty', what do you think it means?" The question is not an open one but a closed one, with one correct answer: loyalty means commitment to the boss, or more to the point, to the organization one is part of. Its interests and values are the "higher good" the individual is obligated to help advance and secure.

But this is a truncated and highly selective understanding of loyalty, bordering on the perverse, in the sense I have set out above: treating something evil/bad as if it is something good. It fails the elementary ethical test: it is not compatible with, nor aligned with, individuality in Alford's sense of autonomy. It is loyalty in a form sacrificing the latter, obliterating it were an individual to dare exhibiting it, insisting on it being the only moral standard that merits compliance. True, individuals who concentrate one-sidedly and uncompromisingly on what their inner tribunal tells them is the right thing to do, may be wrong, even fatally wrong, as Hegel demonstrates in his portrait of the "beautiful soul" (1977 [1807]: 383ff.)). That risk aside, the enduring legacy of individuals like Carson and Thunberg – and I should mention also Rosa Parks here, making it a female trio – is that they not only remind us that organizations belong to "the larger world", but that every individual agent also does so. Hence the good to be aiming for, to help protect, is the common good, not reducible to any one individual's effort to advance it, yet unattainable without it.

4.8 Reconsidering the role of institutions

Thunberg's development as an activist sheds light on some key features in the dynamic between individual and group that merit examination, helping us get a better understanding of what is decisive for resistance to succeed.

As we saw with Georg Seesslen's critique of Thunberg above, she has come under fire for being "divisive" and for deviating from what should be her main – or sole – focus as an activist, namely the climate emergency. Her taking part in pro-Palestinian demonstrations that protest Israel's Netanyahu-led government's atrocities against civilians in Gaza is the most salient example.

How free is a world-known activist like Thunberg to go her own ways, politically and in terms of the protests she takes part in? What it problematic about her no longer being an "innocent" 15-year-old girl? About her becoming more ideologically outspoken, locating the climate emergency within a broader political and economic context, calling what she perceives as the enemies of the required change by name, be it "capitalism" or specific fossil companies? That said, how can organized resistance with regard to the climate issue be effective without being explicitly, perhaps even aggressively, political? If Thunberg as per today, after more than six years of activism, conferences and meetings with world leaders, has come to share the view signalled in Naomi Klein's aptly titled *This Changes Everything. Capitalism vs. the Climate* (2014; Ghosh 2021), a critical analysis and a call to arms wrapped into one, why shouldn't she say so? For fear of losing popular support? Or for fear of attracting ever-fiercer opposition from the most powerful corporations in the world?

To be sure, Thunberg's trajectory from political innocence – "my only message is: listen to the science" – to a politically controversial figure recalls Malm's account of the "radical flank" insofar as she herself now embodies and spearheads such a flank. Recall, also, that Rachel Carson, just like Martin Luther King Jr., became increasingly confrontational in her manner of speaking truth to power, having lost all illusions that change would come about by her being "balanced", polite, and patient. In her speeches in 1963, weakened by the cancer that was to lead to her premature death in 1964, Carson stated her case more and more strongly, expressing her conviction that "no civilization can wage relentless war on life without destroying itself, and without losing the right to be called civilized". In her last public appearances, she focused on "the political and economic forces that prevent changes in pesticide policy", calling out for better legislative control of pesticides no matter how vehemently said forces would try to oppose it, overtly and covertly, and no matter the price she herself would pay when they went after her, trying to silence her (Lear 2009: 439, 440).

In my view, there are important lessons to be learned from Carson's and Thunberg's examples, lessons vital to understanding the pitfalls and traps that a resistance movement needs to be aware of. In his book *Group Psychology and Political Theory* (1994), C. Fred Alford elaborates an analysis of the dynamic between individual and group that is highly relevant to the topic of resistance.

"Most political theory", contends Alford, "is ideology, the ideology of the individual or the ideology of the group. Almost never does it tell the truth, that man [sic] is a group animal at war with his own groupishness". He adds: "Nonetheless, a group of individual members is an ideal, the standard by which political theory should be judged" (1994: 40). Alford cites psychiatrist Vamik Volkan's view that "people need their enemies as much as their allies"; they need, that is, suitable targets for anxieties and bad feelings. Such suitable targets are shared, the group in effect saying to the individual:

> Put your unintegrated and intolerable fear and hate into this group, that symbol. This will relieve your anxiety by allowing you to disown bad parts of yourself, while at the same time allowing you to bond more securely with the good group that is us.

Shared targets, explains Alford, bind groups; it is their glue. "Suitable targets" bridge the distance between individual and group psychology. "Without external enemies, shared targets, the group would fall apart into warring tribes" (41).

Yet the fact that a group needs suitable targets in the sense of "others", *external* to and allegedly very different from the group, is only half the story, albeit often taken as the whole story. The other half is more subtle, but no less potent – or lethal – in its consequences. What is it? It has to do with the fact that the group does not provide the individual member a safe group identity gratis. On the contrary, it demands as payment "that the individual give up his uniqueness to the group, which means that he not express his individuality in the group". This prepares for Alford's major point, insufficiently recognized by Volkan, namely that

> the real threat to the individual is his *own* group, the way in which it demands that the individual suppress his own unique individuality to be a member. It is this threat that is projected into the suitable target and fought there.

In other words, "the real enemy is really one's own group" (42). The upshot is that, to the extent that the critical faculties of its members are "overwhelmed and paralyzed by the anxiety circulating in the group", the thus regressed group "simply can't think". Alford quotes Freud's statement in his classic *Group Psychology and the Analysis of the Ego* to this effect: "The feelings of a group are always very simple and very exaggerated. So that a group knows neither doubt nor certainty" (1959: 10).

As in his study of whistleblowers, when Alford speaks about individuality, he means it in the positive sense of autonomy, key to which is the ability of critical thinking that we just saw risks being suppressed within the group. Yet the individual can only reach the maturity required for such autonomy as a member of a group, in the spirit of Aristotle's argument against self-sufficiency – man is *zoon politikon*. In this sense, "group development is good because, and only because, it allows for individual development. It is individual development that is the highest standard, the *sine qua non* by which group development is judged" (1994: 65). Alford laments that this insight is rarely recognized in the Western canon, For example, for all their differences, Hobbes and Rousseau both get it backwards: "The real work necessary is not to socialize individual men but to individualize social men, to create individuals out of groupies. Groups develop only in interaction with the individuation of their members" (113).

A philosopher more relevant than Hobbes and Rousseau to my discussion about resistance, though, is Rawls, whose account of civil disobedience we examined above. Alford's critique of Rawls' view about the relationship between individual and group is worth quoting *in extenso*:

> In the original position, free and rational persons would agree that there are to be no individuals in the group. My talents, my abilities, my hard work, even my good will and my good character – none of these belong to me in Rawls's scheme. "The assertion that a man deserves the superior character that enables him to make the effort to cultivate his abilities is equally problematic", says Rawls. Everything that I am, all that I have become, even my good character, is regarded solely and strictly as a group resource, to be cultivated only to the degree and extent that it serves the group".
>
> (1994: 136)

What is depicted here is the mentality of the regressed group. Alford admits that, to be fair, Rawls's group is somewhat more advanced – it does not sacrifice difference between individuals for sameness, at least not completely. Differences are acknowledged, says Alford, but they are "severed from real individuals who possess them, as though the existence of embodied individuals with different talents, skills, and personalities is too painful to contemplate" (136). It is as if, for Rawls, it goes without saying that such differences between the individuals making up the group would threaten its ability to function as a cohesive group; and that noticing how others are in some respects – abilities, talents – better than me would inevitably trigger envy rather than admiration or gratitude. Moreover, Alford worries that Rawls seems to suggest that "in the group we attain the excellences that we lack as individuals, because only in the group are we whole". This, says Alford, is "a recipe for the totalitarian group"; indeed, "a recipe for the failed group, one that cannot mobilize the talents and energies of its individuals members as there are not any individual members left" (137); that is to say, not in Alford's sense of autonomy and true individuality. Hence Rawls fails to do justice to the insight Melanie Klein has spelled out, mentioned above: group development is individual development and vice versa. But this does not mean that there is no tension between them, that all is harmony. There is tension, and it is "constant and unremitting. It means that groups that exploit the assets of their members, discouraging their members from appropriating their own talents, are bound to become stultified, reified, and deadened". The upshot is that Rawls gives the lie to his programmatic declaration that his is a "social-contract theory", in that "there are really no individuals anywhere who might debate, discuss, disagree, bargain, and make a contract"; "there exist no individuals to bargain" (138). Importantly, Alford thinks his critique of Rawls applies to Rawls's communitarian critic Michael Sandel as well, regarding them both as exemplifying the trend in political theory to do with its "escape from the real problem of how individuals might live with each other into the metaphysics of abstract individualism or community"; both, that is, "obliterate the individual in the name of a group that doesn't really exist, as it has no real members" (140).

This may strike the reader as a largely theoretical discussion in view of my analysis of resistance as embodied in Carson and Thunberg. What makes it relevant?

The answer comes in several parts. To begin with, recall the observation above that in the current regime of criminalization of activism – not solely in blatantly authoritarian countries like Russia or China, but in the liberal democracies philosophers like Rawls and Habermas have in mind when addressing civil disobedience – a common strategy on the part of authorities is to promote the "othering" of the activists: "they are not like us, not one of us"; "who does she think she is?". In advancing this profile, more is in play than simply encouraging the public to question whether the activists are "representative" – be it of the larger society and the common good, be it of what the science is saying – or if they only pretend to be, as a strategic means to gain support. Driving a wedge between, say, student activists in Extinction Rebellion and "ordinary" working-class people "more worried about paying their bills than about the global temperature some decades ahead" (as the mainstream media have it) is among the oldest tactics of them all, pitting individuals as well as groups against each other to make sure they don't discover that they have common interests. As we saw, the "othering" of Carson as well as Thunberg was (is) richly in evidence, and to some extent it has proven effective with at least portions of the public, though without succeeding in silencing the main target – to the contrary, both Carson and Thunberg were radicalized, becoming more uncompromising, more defiant, by the attempts to isolate and ridicule them. This is part of the dynamic I mentioned above, where tougher methods in targeting activists lead to activists – some of them – resorting to tougher methods in their protests.

Second, Alford's insistence on the importance of affirming difference, as opposed to demanding sameness, between the individuals who make up a group, is absolutely crucial. Note that the tendency to advance sameness at the expense of difference does not come from the group's leader, top down as it were; it comes from *the individuals themselves*, horizontally not vertically, directed at each other to the extent that they – or some of them – do, indeed, come forward as individuals in Alford's emphatic sense. As indicated, to allow for, even actively *cultivate* individuality in the form of one individual's ways of being different from the others – not restricted to talents and abilities, as in Rawls, but including a wider psychological repertoire as well (as with Thunberg's autism) – is instrumental in letting the group develop its full potential, in making it flourish, to use the term from Aristotelian virtue ethics that is so apt here. In this spirit, tension is constant, as Alford says, something to be lived with rather than fought and suppressed. Likewise, disagreement and conflict need to be tolerated as well, not quelled by insisting on consensus, as were it an end in itself and tantamount to cohesion and efficiency – it is not.

What, then, have we learned about the relationship between individual and group that is helpful in trying to understand the potency of resistance – and the obstacles met?

We have seen that according to Alford, the relationship between group (organization) and individual is antithetical, even *antagonistic*, due to "the inability of the organization to tolerate autonomous individuality in the first place", entailing that everybody is "free to participate in a system that is organized to deny the individuality of everyone who gets to play" (2001: 11, 12). So "the organization quells individuality" (21): that is what organizations do.

Is that necessarily so? Is it so in the strong sense of *defining* an organization? I don't think so.

To see why, let me first state the obvious: a study such as Alford's, whose topic is whistleblowing, is all focused on the numerous cases where the story is precisely that of how the individual who blew the whistle is punished for it, the message being: "There is no point in objecting because you are not a person whose exit will be noticed or objection heard. Far from being a hero with a story, you are either a placeholder or patient" (123). Hence, as we have seen, the whistleblower is sacrificed as a lesson to others in the group: "they will see the price of acting as an ethical individual who remembers that he or she belongs to the world" (125). Indeed, this is what Alford was told in his interviews: "When you blow the whistle, you become poison to the company. Your presence makes them sick" (125).

On closer scrutiny, however, this is not the full story, not even in Alford's own books. Recall his statement that "a group of individual members is an ideal, the standard by which political theory should be judged" (1994: 40). It is also, I submit, the standard by which organizations should be judged. This is important to my overall argument about institutions in times of environmental crisis: to avoid throwing the baby out with the water, we need to continue judging institutions – including governments and corporations – by an ideal standard defined by their mission in a modern society: put negatively, to help ensure that as private individuals as well as citizens, we do not act as bastards, as short-sighted egoists indifferent to the greater good; put positively, to help ensure that the sort of individuality Alford considers crucial to demonstrating, embodying and upholding an ethical standard, committed to said common good, be *cultivated* – not quelled – within the institution. Is that too tall an order?

It is not, and it shouldn't be. It is *why* we have schools, universities; why we have law, a parliament and a government, that is, the tripartite separation of powers; why we have professions, like those for doctors, nurses, teachers, architects, lawyers, engineers, businessmen, farmers, fishermen, and carpenters: what happens when the individual enters and becomes part of his or her appropriate institution qua professional is that he or she is taught by experienced role models, by way of example, about the *standards of excellence* that apply to the work to be done there; about teachers', nurses', engineers', etc. commitment to *the good* with which the profession in question is trusted – trusted by the society at large, as *our* good – learning, health – and not that of the individual taken alone. To put it like this is to frame the issue in the way Aristotle and Hegel do, based on the insight that groups make us human, facilitate individuation and autonomy, and promote the moral maturation of the individual member; in short, groups help evolve private individuals into citizens, political beings. In real life, of course, this is never straightforward, never without tension and conflict, and often the suppression of individuality – not its cultivation – is what prevails, thus perverting the mandate given by society. But it *need not* be like that, and to critique a given institution on this score is impossible without holding it to the standard spelled out.

More precisely, what does this view of groups in general and institutions in particular imply for our above discussion about resistance? First, it suggests how to fill the lacuna I spotted in Butler as well as in Malm, that is, their tendency to *dismiss any positive role for existing institutions*, citing the many reasons to distrust them in an era when they have been co-opted by neoliberal reforms. Resulting from this dismissal is a simplified framework focused on the conflictual and highly asymmetric relationship between the state and the big corporations on the one hand and groups of activists on the other (be it Extinction Rebellion, be it Black Lives Matter), with nothing in between, as it were: no agential role is accorded to civil society, to the public, or – to get to my specific point – to the cluster of institutions, organizations, and professions. My Aristotle- and Hegel-inspired argument is that in giving up on the potential for loyalty to the greater good to be promoted as part of the culture of these bodies, critical theorists like Butler and Malm give up on an indispensable agential source of resistance – and for resistance to be effective, it needs to be organized, to be aligned with and anchored in social bodies over time, so as to *persist* rather than evaporate the moment today's demo is over and its individual participants disperse and go home.

It may be objected that this argument is unrealistic, because institutions and professions all across the board have been so thoroughly contained and co-opted by the depoliticized culture of neoliberalism. In my view, despite the considerable degree of truth, this is too defensive. The fight to ensure that loyalty is not to the boss, or to the twin-imperatives of growth and profit, but to the standards of excellence of the profession/institution in question, is a fight that isn't simply over, a thing of the past, and settled once and for all, but *ongoing* in so many places, including my own, the university; under Trump 2. 0, the stakes involved have become immensely higher in a matter of months, threatening to forbid the topics that make up a book such as this one, if not by law, then by other means aimed at their silencing. This means that, as far as Alford's theory is concerned, for all its valuable insights, it is too individualistic: it is so disillusioned with how organizations "typically" treat their truly individual member, based on their rise and fall as whistleblowers, as to virtually give up the standard Alford himself at one crucial point says we must hold them to. The relationship between organization and individual *cannot* be of the zero-sum kind, where what empowers the one crushes the other. Instead it must be held to a standard insisting on the potential for mutual maturation and support.

4.9 Leadership in troubled times

Since stories of the resistance whistleblowers attract seem to be everywhere these days, and political breakthroughs few and far between, I shall end this discussion on a more constructive and forward-looking note. I will do so by moving from the relationship between individual and group (or society at large) to that between leader and group. It suits my purpose to use an example of leadership – in the positive sense of the role – employed by Alford himself, illustrating many of the points discussed above. Showing the importance of leadership is all the more important since it is a topic regrettably neglected in much leftist discussion about resistance, and why it so often fails and falls apart. The fact that leadership is largely absent in Butler's work, and not systematically a part of Malm's agenda, echoes among critical theorists the conspicuous absence of bold, enduring and memorable leader figures in movements like Occupy Wall Street, Black Lives Matter, and Extinction Rebellion – as if leadership were a conservative topic and role, epitomized, say, by Winston Churchill, not required for revolutionary change, and as if "flat structure" is superior to any form of hierarchy.

In my view, leaders who come forward and lead by example are indispensable to the political struggle for a better society. The example Alford examines is Vaclav Havel and his "New Year's Address" of 1990, the first of his presidency in what was recent post-Communist Czechoslovakia at the time. In what follows I quote the crucial points made by Havel, and Alford's comments on them, summing up much of the discussion of this chapter.

Havel starts his speech in this way: "I assume you did not propose me for this office so that I, too, would lie to you". Alford comments: "Probably nothing is more important than this as far as good leadership is concerned, because it helps citizens and members integrate their split perceptions of the good and the bad leader".

Havel's second point is this: "Our country is not flourishing". He relates how he recently flew over Bratislava and was appalled at the ugliness and the pollution: "This view was enough for me to understand that for decades our statesmen and politicians did not look or did not want to look out of the windows of their airplanes".

His third point: "The worst thing is that we live in a contaminated moral environment. We fell mortally ill because we became used to saying something different from what we thought. We learned not to believe in anything". Alford comments:

> The illness stems from within, not without. We have ourselves to blame, the way we split ourselves off from reality". Havel goes on: "We have to accept this legacy as a sin we committed against ourselves. If we accept it as such, we will understand that it is up to us all, and up to us only, to do something about it. We cannot blame the previous rulers for everything.

Comments Alford: "Paranoid projection, scapegoating, the sacrifice of failed leaders, the discovery of new enemies: all this is contained, rather than encouraged, by Havel".

His fourth point: "But if the responsibility stems from within, so too will the solution. Freedom and democracy include participation and therefore responsibility from us all".

His fifth point: "Only if we accept both truths – that the responsibility lies within, but so too does the solution – will hope return to our hearts". Alford comments:

> Overcoming depressive anxiety that we are not strong enough or good enough to repair the damage we have done by our own greed and fear requires that we first confront our responsibility for making things the way they are. Only then is genuine hope, as opposed to manic denial, possible. Depressive anxiety is the source of corruption. [...] But even this hope is not free. Gaining access to it requires that we confront our responsibility not only for our current environmental and moral pollution but also for the suffering of those who all along tried to speak the truth and were persecuted for it. There was a part of our group that always spoke the truth and was sacrificed for it. We must not forget them, for they represent, or held, a part of ourselves that knew the truth but would not speak it. Only by reclaiming this part can we go on.

We may think of Carson and Thunberg when reading this.

His sixth point:

> If the source of hope and restoration is within, so too is the enemy. Our main enemy today is our own bad traits: indifference to the common good; vanity; personal ambition; selfishness; and rivalry. The main struggle will have to be fought on this field.

Alford comments:

> To this I would add that the main enemies are fear and despair, the depressive anxiety that we are not good enough or strong enough or adult enough to face reality and change it. It is in the face of this anxiety that corruption, of ourselves, our leaders, and our relationships, seems the only alternative.
>
> (1994: 164, 165)

Alford sums up:

> Without a good leader citizens are unlikely to reclaim enough of themselves from the group to relate to each other as individual members. Every step in Havel's story is about encouraging citizens to reclaim both their fears and their capacity to confront them. Every step rejects the collusion between leader and led, the charms of corruption. [...] What makes the difference between a good and a bad leader is whether the leader promotes unification by raising the ego ideal, while encouraging and supporting the members in their efforts to live up to it. Or whether the leader promotes unification by lowering the ego ideal and so telling members in effect that they are already perfect, having no need to develop and grow up. The latter strategy is corruption, the paradigm of all corruption, in which the importance of work, effort, maturity, and time to the achievement of valued goals is denied. [...] The demagogue appeals by telling members that they are already perfect, if only they knew it or if only some enemy did not stand in the way of reunification with perfection. Often this enemy is characterized as polluted, dirty, or in some way diseased, so that perfection becomes tantamount to cleansing.
>
> (1994: 166, 167)

Indeed, the difference between leadership á la Havel and leadership á la Trump could hardly have been greater, being worlds apart in substance as well as in form. It is often said that, at least in democracies, people get the leaders they (we) deserve. And to be sure, there is a sense in which the re-election of Trump is a mirror image of pervasive trends in the American society, trends increasingly manifest in many European countries as well, as witnessed in the rise of right-wing populism in recent years.

Crucial among these trends is the hegemony of our culture's "first language" of utilitarian individualism, of "what's in it for me?", a hegemony making for a throughgoing linguistic and moral impoverishment. As Anna L. Peterson, an American professor of religion, has noted, "because this first language so powerfully shapes the way we think, even non-individualistic actions are filtered through the lens of individualism" (2009: 125). She mentions a striking example: asking students in an environmental ethics class if they could think of something they had done for reasons other than self-interest, none of the thirty-five students, many of whom were volunteers and activists, listed a non-self-interested reason for doing good. They described their altruistic practices as motivated by their interest in self-esteem or good company; and they did not connect "feeling god" about doing good things to others and for the community "to deeply held, non-instrumental values" (2009: 125). To put it in a way not employed by Peterson, but in keeping with the topic of this book, the students demonstrate what it means to be *in denial about the human capacity for doing good*.

Putting it like this allows me to make a final point about perversion, one not made when I discussed perversion at a structural, as distinct from individual-oriented level in Chapter 1 – namely, the perversion in play in refusing to recognize something as *good* even when staring it in the face, be it in the form of some act done by others, or some act I myself perform. That doing good, and being motivated to do good, for the sake of those to whom that action is directed, is nowadays a notion held in suspicion, even when it's based on personal experience, speaks volumes about the perversion, the turning of tables, involved when distrust comes to replace trust as the shared, quasi "given" presupposition in our social interactions, as well as in institutions and in politics. Giving up on the reality, or even the *possibility* of the good, its likelihood to stand a chance in a world in the grips of multiple crisis, may well prove to be the most consequential triumph of Trumpism, with repercussions far beyond the United States.

The stakes could scarcely be higher, since the alternative to a leadership, such as Havel's, that seeks to acknowledge the capacity for the good *as well* as for evil in each of us, in every individual as well as in every group, in "us" as well as in "them", risks being that of antagonism as settled only by violence, whereby violence springs from the conviction that the other is our enemy. At its core, and as pointed out by the Hegel-inspired Bosnian philosopher and theologian Mile Babic,

> violence springs from our need for self-aggrandizement, self-congratulation and self-worship. Violent people and collectives fail to perceive that in destroying others [including so many other-than-human others – AJV] they are destroying themselves. This is the blindness, the sheer stupidity of violence, or to put it better, the pathology of violence.
>
> (Babic 2019: 206)

The sort of leadership needed today is that of warning against the sources, internal as well as external, of the violence-promoting alternative to a belief in the possibility of goodness.

Bibliography

Alford, C. Fred, 1994. *Group Psychology and Political Theory*. New Haven: Yale University Press.
Alford, C. Fred, 2001. *Broken Lives and Organizational Power*. Ithaca: Cornell University Press.
Arendt, Hannah, 1978. *The Life of the Mind*. London: Secker & Warburg.
Babic, Mile, 2019. *The Violence of the Idols*. Sarajevo: University Press.
Baudrillard, Jean, 1996 [1968]. *The System of Objects*. London: Verso.
Baudrillard, Jean, 1998 [1970]. *The Consumer Society*. London: Sage.
Bauman, Zygmunt, 1989. *Modernity and the Holocaust*. Cambridge: Polity Press.
Bauman, Zygmunt, 2007. *Consuming Life*. Cambridge: Polity Press.
Bauman, Zygmunt, 2008. *Does Ethics Have a Chance in a World of Consumers?* Cambridge, MA: Harvard University Press.
Bauman, Zygmunt, 2011. *Collateral Damage. Social Inequalities in a Global Age*. Cambridge: Polity Press.
Beckert, Jens, 2024. *Verkaufte Zukunft. Warum der Kampg gegen den Klimawandel zu scheitern droht*. Berlin: Suhrkamp.
Brennan, Teresa, 1993. *History After Lacan*. London: Routledge.
Brennan, Teresa, 2003. *Globalization and its Terrors*. London: Routledge.
Callicott, J. Baird, 1999. *Beyond the Land Ethic*. New York: SUNY.
Ernman, Malena, 2019. *Huset brenner*. Oslo: Cappelen Damm.
Farrell, Clare, et al. Eds., 2019. *This is Not a Drill. An Extinction Rebellion Handbook*. New York: Penguin.
Foster, John Bellamy, 2000. *Marx's Ecology*. New York: Monthly Review Press.
Foster, John Bellamy, and Fred Magdoff, 2011. *What Every Environmentalist Needs to Know About Capitalism*. New York: Monthly Review Press.
Freud, Sigmund, 1985 [1921]. "Group Psychology and the Analysis of the Ego", in *Civilization, Society, and Religion*, 95–178. Harmondsworth: Penguin.
Ghosh, Amitav, 2021. *The Nutmeg's Curse*. London: John Murray.
Giddens, Anthony, 2011. *The Politics of Climate Change*. Cambridge: Polity Press.
Girard, Rene, 1988. *Violence and the Sacred*. London: Athlone Press.
Habermas, Jürgen, 1990. *Moral Consciousness and Communicative Action*. Cambridge: Polity Press.
Habermas, Jürgen, 1996. *Between Facts and Norms*. Cambridge, MA: MIT Press.
Hallam, Roger, 2019. "The Civil Resistance Model", in Farrell, Clare, et al., eds., *This is not a Drill. An Extinction Rebellion Handbook*. London: Penguin.
Harvey, David, 2010. *The Enigma of Capital and the Crises of Capitalism*. London: Profile Books.
Harvey, David, 2014. *Seventeen Contradictions and the End of Capitalism*. London: Profile Books.
Havel, Vaclav, 1983. *Briefe an Olga*. Hamburg: Rowohlt Verlag.
Hegel, Georg F. W., 1978 1907]. *The Phenomenology of Spirit*. Oxford: Oxford University Press.
Hegel, Georg F. W., 1991 [1921]. *Elements of the Philosophy of Right*. Cambridge: Cambridge University Press.

Hochschild, Arlie Russell, 2016. *Stranger sin Their Own Land. Anger and Mourning on the American Right*. New York: The New Press.
Hochschild, Arlie Russell, 2024. *Stolen Pride. Loss, Shame, and the Rise of the Right*. New York: The New Press.
Hornborg, Alf, 2019. *Nature, Society, and Justice in the Anthropocene*. Cambridge: Cambridge University Press.
Hylland Eriksen, Thomas, 2016. *Overheating. An Anthropology of Accelerated Change*. London: Pluto.
Hylland Eriksen, Thomas, 2018. *Boomtown. Runaway Globalisation on the Queensland Coast*. London: Pluto.
Jackson, Tim, 2010. *Prosperity without Growth*. London: Earthscan.
Klein, Melanie, 1989. *Love, Guilt, and Reparation*. London: Virago.
Klein, Naomi, 2014. *This Changes Everything. Capitalism vs. the Climate*. London: Allen Lane.
Kovel, Joel, 2007. *The Enemy of Nature*. London: Zed Books.
Krznaric, Roman, 2020. *The Good Ancestor*. London: W. H. Allen.
Lear, Linda, 2009. *Rachel Carson. Witness for Nature*. New York: Harcourt.
Lysaker, Odin, 2024. *Ecological Democracy. Caring for the Earth in the Anthropocene*. London: Routledge.
Malm, Andreas, 2016. *Fossil Capital. The Rise of Steam Power and the Roots of Global Warming*. New York: Verso.
Malm, Andreas, 2018. *The Progress of This Storm*. New York: Verso.
Malm, Andreas, 2021. *How to Blow Up a Pipeline. Learning to fight in a World on Fire*. New York: Verso.
Malm, Andreas, and Wim Carton, 2024. *Overshoot. How the world Surrendered to Climate Breakdown*. New York: Verso.
Monbiot, George, 2024. "It's a plutocrat's world – and all dissenters are swiftly crushed", in *The Guardian Weekly* 9 February, 46.
Næss, Petter, and Leigh Price, 2016. *Crisis System*. London: Routledge.
O'Connor, James, 1998. Natural Causes. *Essays in Ecological Marxism*. New York: The Guilford Press.
Peterson, Anna L., 2009. *Everyday Ethics and Social Change*. New York: Colombia University Press.
Rawls, John, 1971. *A Theory of Justice*. Oxford: Oxford University Press.
Rolston, Holmes, 1988. *Environmental Ethics*. Philadelphia: Temple University Press.
Saito, Kohei, 2022. *Marx in the Anthropocene. Towards the Idea of Degrowth Communism*. Cambridge: Cambridge University Press.
Seesslen, Georg, 2024. "Das Märchen ist zu Ende", in *Der Freitag 24*. Oktober.
Thunberg, Greta, 2019. *No One Is too Small to Make a Difference*. London: Penguin.
Vetlesen, Arne Johan, 2015. *The Denial of Nature*. London: Routledge.
Vetlesen, Arne Johan, 2016 "*Making the critique of capitalism fit for the present Crisis*", in Naess and Leigh, eds., 209–224.
Vetlesen, Arne Johan, and Rasmus Willig, 2018. *Hva skal vi svare våre barn?* Oslo: Dreyer.
Vetlesen, Arne Johan, 2019. *Cosmologies of the Anthropocene*. London: Routledge.
Vetlesen, Arne Johan, 2023a. "Ethics in the Anthropocene: The case for questioning anthropocentrism", in *Constellations* 30 (2), 153–161.
Vetlesen, Arne Johan, 2023b. *Animal Lives and Why They Matter*. London: Routledge.
Weis, Valeria Veg, ed., 2022. *Criminalization of Activism*. New York: Routledge.

Index

Kolibri 360 GmbH
Ferdinand-Juhlke-Straße 7
99095 Erfurt, Deutschland
produktsicherheit@kolibri360.de